**Name:**

**Email:**

**These are my class notes, and I really need them to study. If you find this book, please contact me or leave the book in the classroom where I can find it. Thank you!**

ORGANIC CHEMISTRY 1 LECTURE GUIDE 2015

BY RHETT C. SMITH, PH.D.


Marketed by Ideal Quill Publishing Group, LLC


Printed in the United States of America

10 9 8 7 6 5 4 3 2 1

ISBN 978-0692499924

# Organic Chemistry 1 Lecture Guide 2015

**Rhett C. Smith, Ph.D.**

# Contents

# Lecture Set 1:
# Introduction to Organic Chemistry

**Suggested Reading:**

**Suggested Problems:**

# Rules of the Game

## 1. Opposite Charges Attract

## 2. Like Charges Repel

**Notes:**

# Review: Informal Definitions I

• Valence Electrons:

• Electronic Configuration:

• Lewis Dot Structure:

The Periodic Table gives you:

The number of unpaired electrons in the **Lewis Structure** tells you:

**Notes:**

# Lewis Dot Structures

**When given the formula for a molecule or ion:**

1. Add up valence electrons; if it is an ion and has a charge, add an electron for each "–" charge, take away an electron for each "+" charge.

2. Draw single bonds between the atoms

3. Add remaining valence electrons to the atoms to give them the expected number of lone pairs (you know this from the Lewis structure of the given element)

4. Add bonds between adjacent elements that have open valences, making multiple bonds.

**HINT:**

Most neutral species we will encounter in this class feature elements with a total of 8 electrons (an **octet**) around them in the correct structure, except H, which will have only 2.

**Notes:**

# Lewis Dot Structure Examples

**Target** **Atoms** **# electrons** **structure**

(A) $CCl_4$

= total

(B) $CH_3OH$

= total

(C) $CH_2O$

= total

**Notes:**

# TIMESAVERS: Shorthand Representations

| dash formula | condensed formula | bond-line formula |
|---|---|---|
| (A) 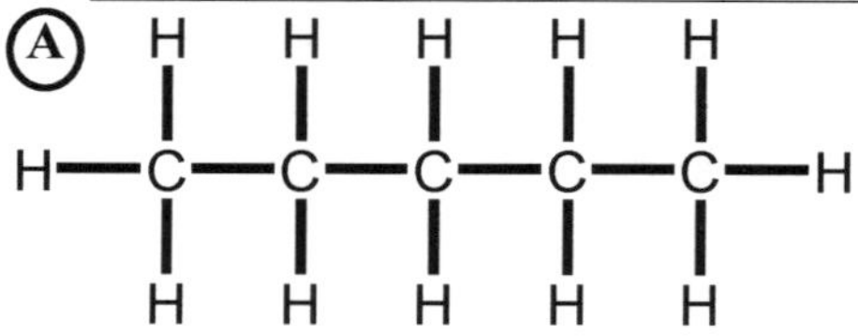 | $CH_3CH_2CH_2CH_2CH_3$<br>$CH_3(CH_2)_3CH_3$ | |
| (B) 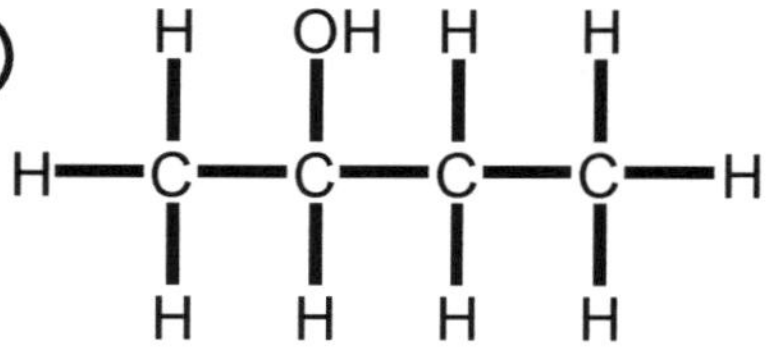 | $CH_3CH(O)CH_2CH_3$ | |
| (C) 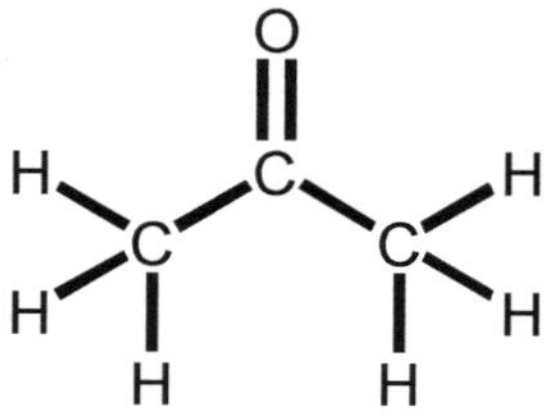 | $CH_3C(O)CH_3$ | |

**WE WILL USE THESE FROM NOW ON**

**Notes:**

# Electrons Between Nuclei Make Bonds

**ionic bonds**: "Complete" transfer of electrons to give two ions

**cation** **anion**

**covalent bonds**: Equal or Unequal sharing of electrons without "complete" transfer

A bond can be represented by two dots or by a line.

**A line represents two electrons!**

**The extent to which the electron is transferred depends upon how strongly each atom attracts the electrons within the bond. This attractive force is called the ELECTRONEGATIVITY of the atoms.**

**Electronegativity (EN):**______________________________.

*increases______________________________in the periodic table*

**Notes:**

# Electronegativities

The difference in electronegativity (EN) of two bonded atoms will tell you to what extent the electrons in the bond are transferred from one atom to the other.

H—H
EN = 2.1 2.1

Cl—Cl
EN = 3.0 3.0

} $\Delta EN = 0$ equal sharing of electrons
= ______________________________.

$\delta^+$ $\delta^-$
H—Cl
EN = 2.1 3.0

$\Delta EN = 0.9$ unequal sharing of electrons
= ______________________________.

$Na^+Cl^-$
EN = 0.9 3.0

$\Delta EN = 2.1$ transfer of electrons
= ______________________________.

**general rule: $\Delta EN$ < ____. = "covalent"**
**> ____. = "ionic"**

**Δ = uppercase "delta". In science and math, delta often represents a change or difference.**

**Notes:**

# Representing Polar Bonds

$\delta^+$, $\delta^-$

**These are lowercase "delta" symbols. In science and math, delta often represents a change or a difference.**

**A) ______________________ : these lowercase Greek symbols ('delta') tell us that the atom above which they are written bear a partial positive or negative charge.**

$\delta^+$ $\delta^-$

**H—F**

**The dipole moment is represented by 'mu', the greek 'm'.**

**B) An arrow can also be used to represent the pull of electrons in the direction of the arrow. To emphasize the direction of electron movement, a 'plus' sign is drawn at the other side of the arrow. These arrows show us the ______________________ ($\mu$) of the molecule.**

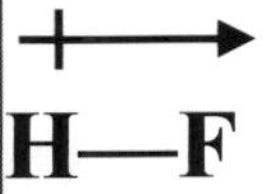

## Notes:

# Dipole Moments

Draw in the dipole moments for these molecules and determine which has the largest dipole moment in each set.

**Set 1:** $CCl_4$, $CO_2$, $CHCl_3$, $CF_4$

**Set 2:** $CH_4$, $CH_3Cl$, $CH_2Cl_2$, $CHCl_3$

**Notes:**

# Where is the Attraction?

Indicate the place(s) on each molecule to which an anion or electron pair would be attracted:

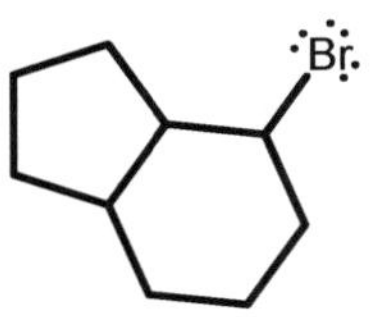


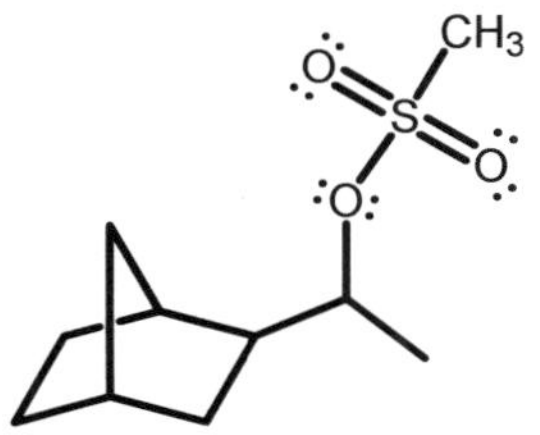


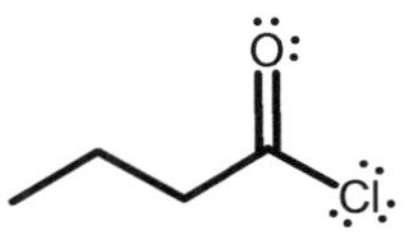


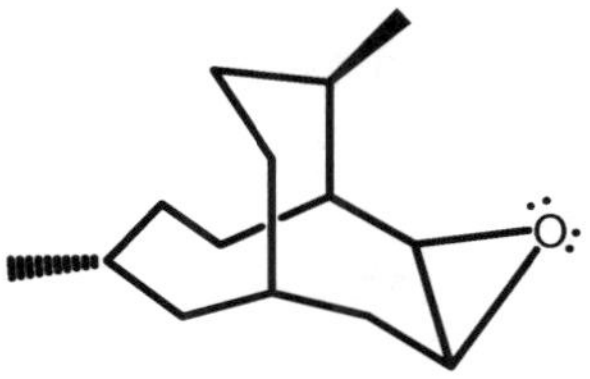


Indicate the place(s) on each molecule to which a cation or other electron-deficient species would be attracted:

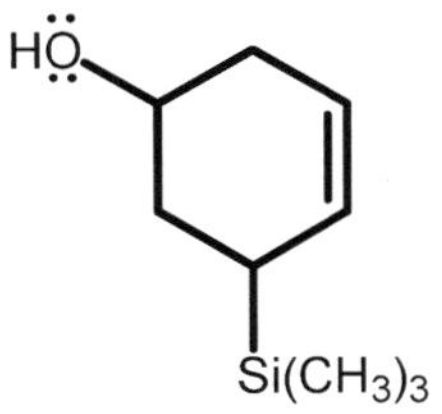


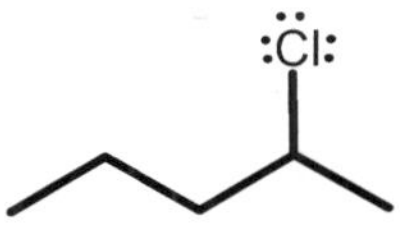


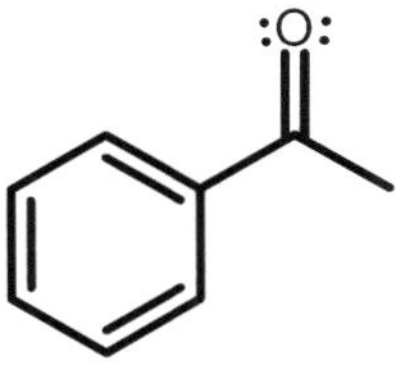


## Notes:

# Intermolecular Forces

The attraction between positive and negative charges also plays a role in the properties of molecules. A few important intermolecular forces are reviewed below:

**Weaker**

van der Waals forces/London dispersion forces:

Dipole-dipole interactions:

Hydrogen bonding:

**Stronger**

## Notes:

# Boiling Points

Molecules with stronger intermolecular forces will tend to more strongly hold molecules together in a condensed state (solid or liquid). Molecules with weaker intermolecular forces will be broken apart into the vapor phase, where molecules are farther apart from one another, at a lower temperature. This lets us predict boiling points on the basis of intermolecular force strength.

Rank each set of molecules in terms of boiling point, high to low:

**Set 1:**

$CH_3CH_2CH_2CH_2CH_3$

$CH_3CH_2CH_2CH_2CH_2OH$

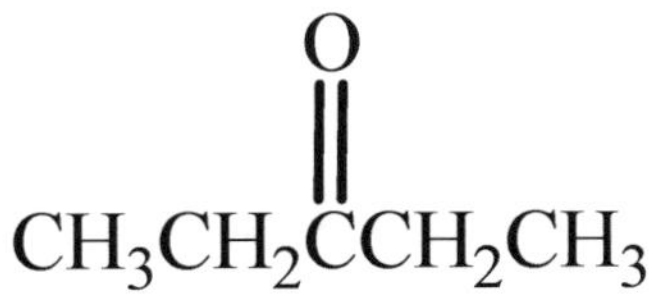

**Set 2:**

$CH_3CH_2CH_2CH_3$

$CH_3CH_2CH_2CH_2Cl$

$HOCH_2CH_2CH_2CH_3$

$HOCHCH_2CH_2CH_2OH$

**Notes:**

# Representing Transfer of Electrons: "Arrow Pushing"

**Since bonds consist of electrons between nuclei, the movement of electrons leads to changes in bonding; in other words Chemical Reactions are described by showing how the electrons rearrange upon going from reactants to products. In order to show the movement of electrons, we need a notation that is clear.**

**Most of the time we will deal with electron pairs moving (recall that a bond is made of a pair of electrons!). We use a simple curved arrow to show the direction of electron.**

**In those less frequent cases (in this class) when we will move only ONE electron, we use a 'fishhook' arrow; that is, an arrow with only a half arrowhead.**

**Notes:**

# Arrow Pushing

**Electrons will move FROM a good electron source (for example a lone pair or an anion, or the negative end of a polar bond ...) TO a good electron acceptor or "sink" (which may be the positive end of a polar bond, an empty orbital, a cation ...).**

**Most of this class will deal with determining how electrons move, rationalizing why they do so, and predicting how they will move in new circumstances. In other words, this is a very important portion of the notes to study!**

**Notes:**

# Nucleophiles and Electrophiles

NUCLEOPHILE:

ELECTROPHILE:

**Notes:**

# Reaction Steps

In organic chemistry, we look at quite a few reaction mechanisms, some of which involve many individual steps to get from starting materials to the final product. It is worth noting that many reactions, even rather complicated ones, are composed predominantly of a few basic general mechanistic steps:

**1. Coordination/Heterolysis (Lewis Acid / Base reaction):**

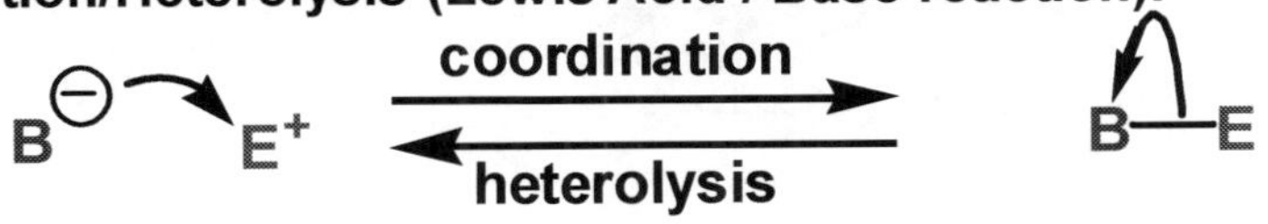


**2. E2 Reaction:**

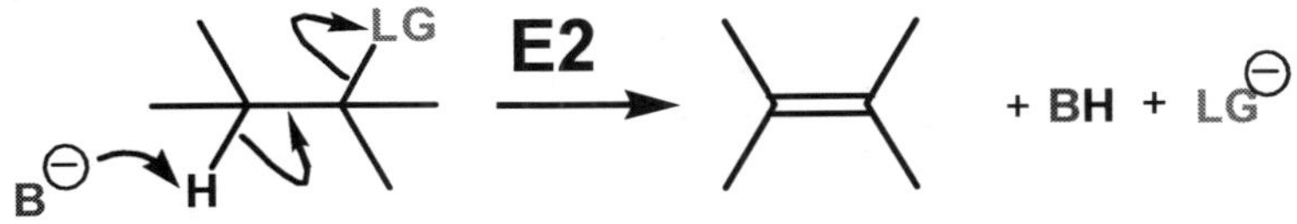


**3. Electrophilic Addition / Elimination:**

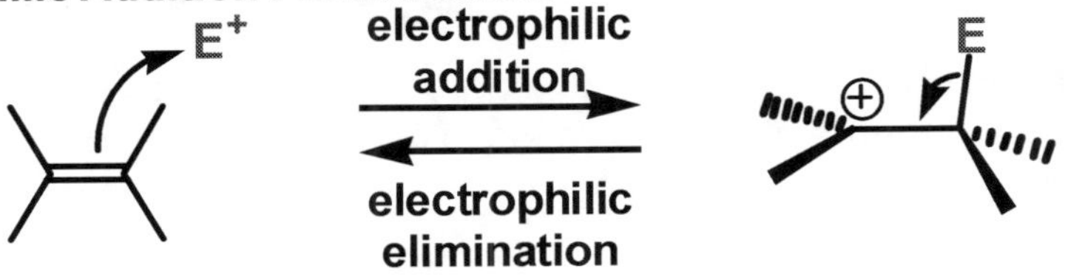


**Notes:**

# Reaction Steps

**4. Carbocation Rearrangement**

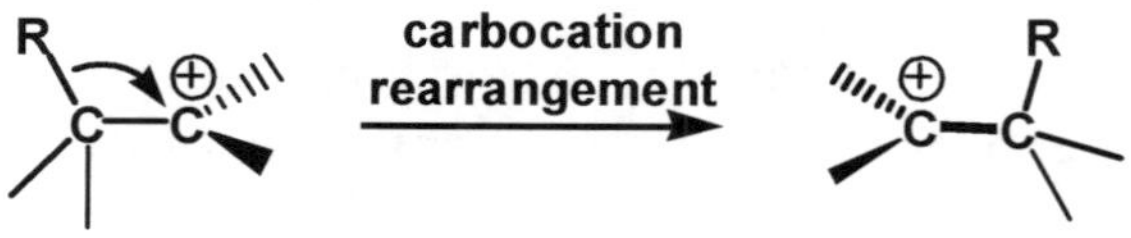


**5. SN2**

**6. Nucleophilic Addition/Elimination**

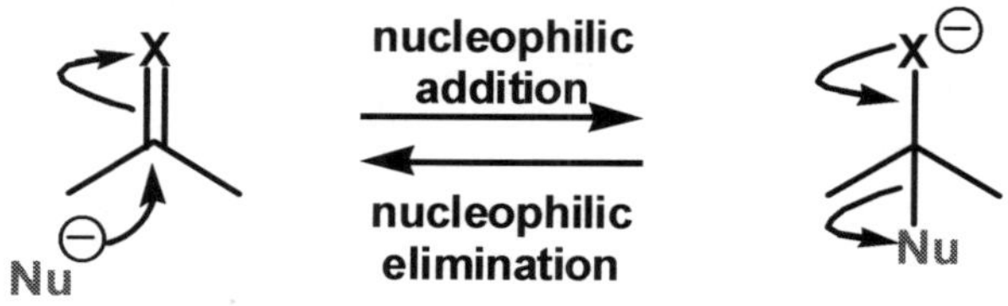


When designing or analyzing a reaction, evaluate the driving force for each step; electrons tend to move from electron rich sites to electron poor sites in order to reach more stable species.

**Notes:**

## Formal Charges

Definition:

Find the formal charge for each of the indicated elements for the species shown:

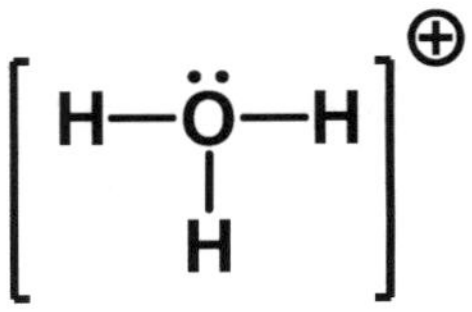

O:

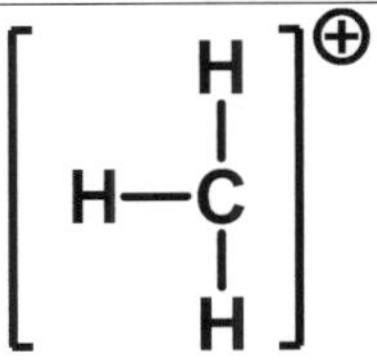

C:

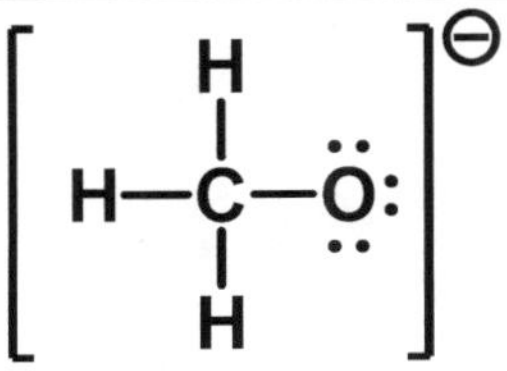

C:
O:

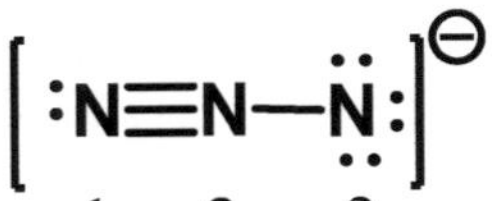

1 2 3

N#1:
N#2:
N#3:

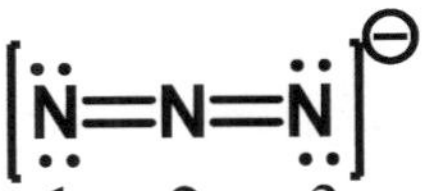

1 2 3

N#1:
N#2:
N#3:

C:
O:

**Notes:**

# Resonance Structures

In a resonance structure the atomic positions (connectivity) remain the same, but electrons move around (lone pairs and multiple bond positions change).

neither resonance form alone accurately describes the anion; the "real" structure is a combination of the two, as shown at right

(dashed line represents a partial bond)

**Why do we care?** Well, sometimes resonance structures help us understand reactivity. Also, if a molecule can be drawn in numerous resonance forms, it has some added stability known as "**resonance stabilization**". This extra stability is due to: **delocalization** of electrons and/or charge=

**Notes:**

# Resonance Structures

1. Draw the resonance hybrid for **Structure A**, which could form upon protonation of the oxygen as shown:

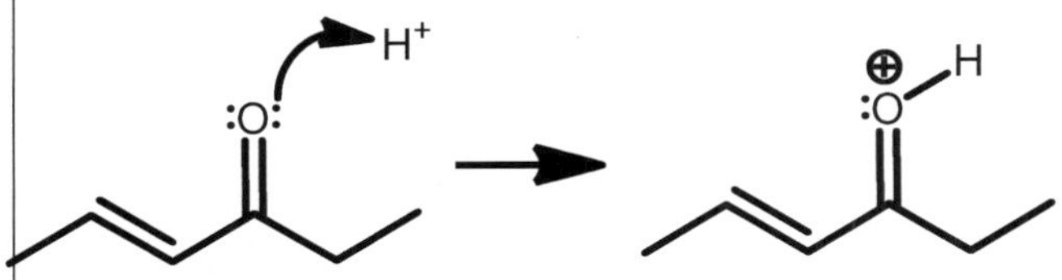


Structure A

A

2. Which is **not** a resonance structure for the molecule shown?

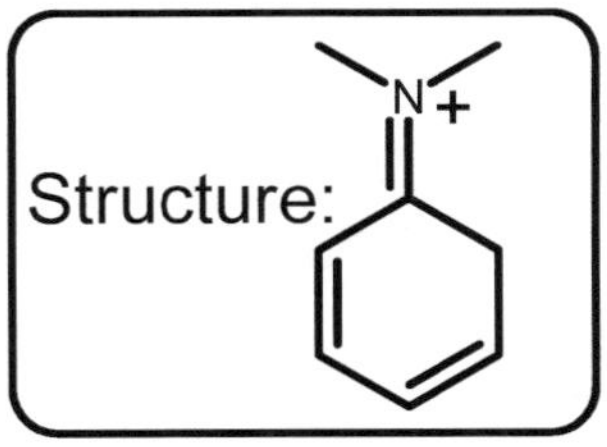


(A)

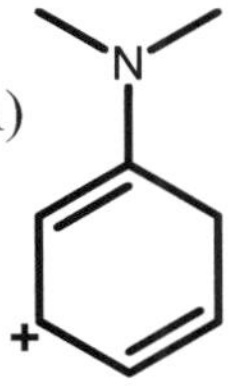


(B)

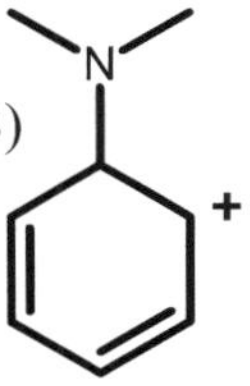


(C)

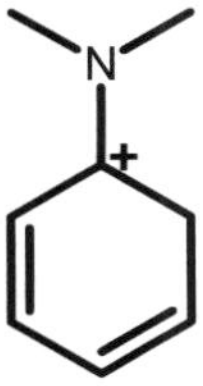


(D)

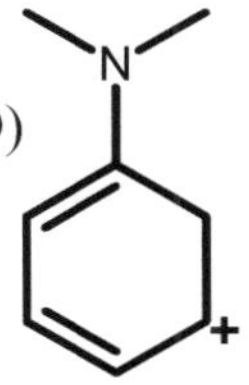


**Notes:**

# Arrows in Organic Chemistry

**Notes:**

# Where are the electrons? … Orbitals.

| *l* | 0 | 1 | 2 | 3 | 4 | 5 |
|---|---|---|---|---|---|---|
| Name | s | p | d | f | g | h |
| Shapes | 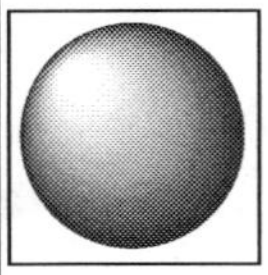 | 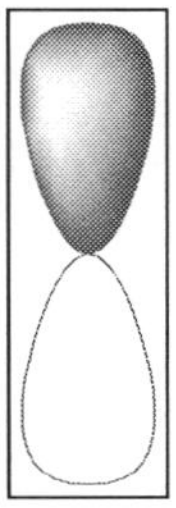 | 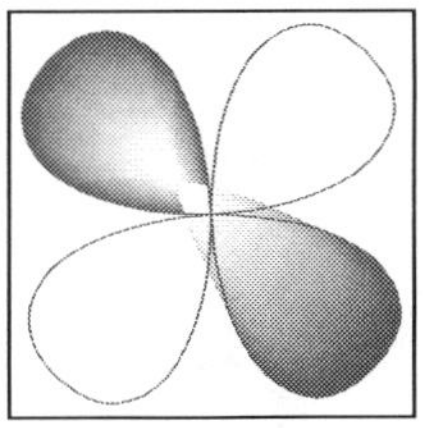 | | | |

**s orbitals are spherical**

**p orbitals are hourglassical**
**There are three p orbitals in each shell;**
**One on each of the axes (x, y, and z)**

**"Hourglassical" nomenclature courtesy of Andrew G. Tennyson, Professor Extraordinaire**

## Notes:

# Showing a Molecule's 3D shape on 2D Paper

**Since orbitals point in different directions, it should not be surprising that bonds point in various directions, and that molecules are three-dimensional. Chemists use a simple shorthand, called the "dash-wedge notation" in order to attempt to indicate a molecules 3D shape on 2D paper. It's not too complicated ...**

**A "regular" line indicates a bond that lies**

(A)

**A "wedge" indicates a bond that is**

(B)

**A "dashed wedge" indicates a bond that is**

(C)

**Notes:**

# Hybridization and Molecular Orbitals

**Two Orbitals can combine ...**

**recommended:**
**You should be able to sketch an orbital resulting from the addition of any two orbitals given.**

... by adding ...

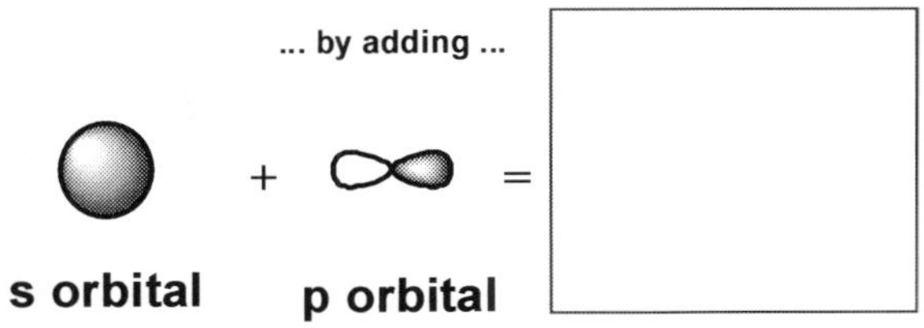


... or by subtracting ...

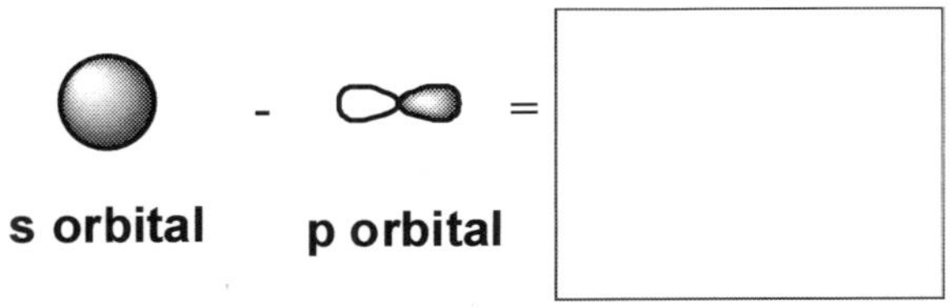


**... to form two new orbitals.**

Adding or "hybridizing" one s orbital with one p orbital gives two **sp hybrid orbitals**.

**(not to scale)**

# Notes:

# The sp hybridized atom

**Drawing both sp hybrid orbitals on the same atom:**

This is an **sp hybridized** carbon atom. The sp orbitals are pointed directionally [ ] from one another, in what is known as a **linear geometry**.
Since carbon has three p orbitals and we only used one to make the hybrids, two p orbital are still left unhybridized on carbon.

The remaining p orbitals lie perpendicular to the line of the sp hybrid orbitals.

Using the dash-wedge notation, we can denote the geometry and show the p orbital that remains unhybridized.

## Notes:

# The $sp^2$ hybridized atom

**Drawing all three $sp^2$ hybrid orbitals on the same atom:**

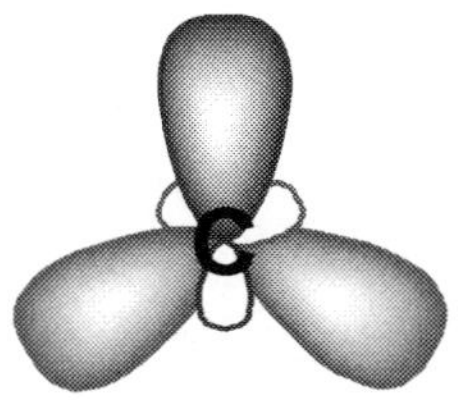


This is an **$sp^2$ hybridized** carbon atom. The $sp^2$ orbitals are in a plane pointed directionally______° from one another, towards the corners of an equilateral triangle, in what is known as a _______________________________geometry.

Since carbon has three p orbitals and we only used two to make the hybrids, a p orbital is still left unhybridized on carbon.

It lies perpendicular to the trigonal plane of the three $sp^2$ hybrid orbitals.

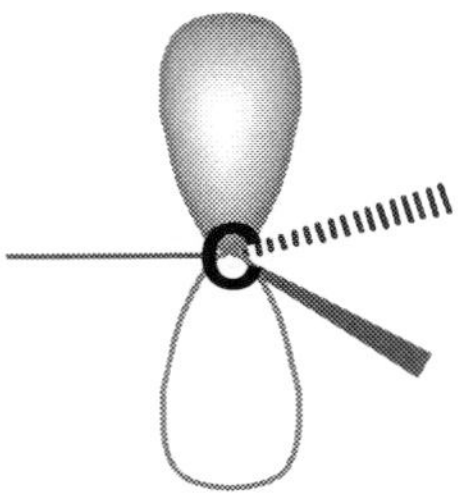


Using the dash-wedge notation, we can denote the geometry and show the p orbital that remains unhybridized.

Hybrid orbitals are involved in forming, or may be filled by a lone pair of electrons.

**Notes:**

# The $sp^3$ hybridized atom

**Drawing all four $sp^3$ hybrid orbitals on the same atom:**

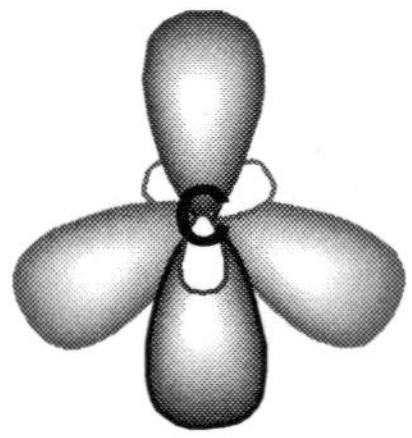

This is an **$sp^3$ hybridized** carbon atom. The $sp^3$ orbitals are pointed directionally _______ ° from one another, towards the corners of a tetrahedron, in a ________________ geometry.

Since carbon has three p orbitals and we used all three to make the hybrids, all of the valence orbitals are now $sp^3$ orbitals, and consequently these are the only ones used in bonding.

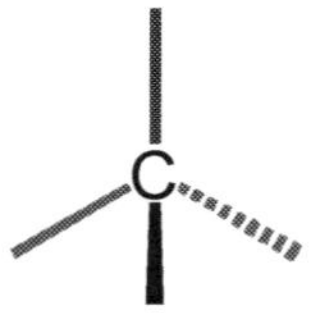


Using the dash-wedge notation, we can denote the geometry and show the p orbital that remains unhybridized.

******************

Hybrid orbitals are involved in forming [ ]

, or may be filled by [ ]

**Notes:**

# Sigma (σ) and Pi (π) Bonds

A sigma (σ) bond is one in which the electrons making the bond are between the two nuclei joined by the bond:

A pi (π) bond is one in which the electrons lie above and below the line between the two nuclei which are joined by the bond:

In the vast majority of example we will study, the **hybrid orbitals** will form sigma bonds or will be filled by a lone pair of electrons, and any pi bonds will be made by p orbitals not hybridized into sp, $sp^2$, or $sp^3$ orbitals…

**Notes:**

# Sigma ($\sigma$) and Pi ($\pi$) Bonds

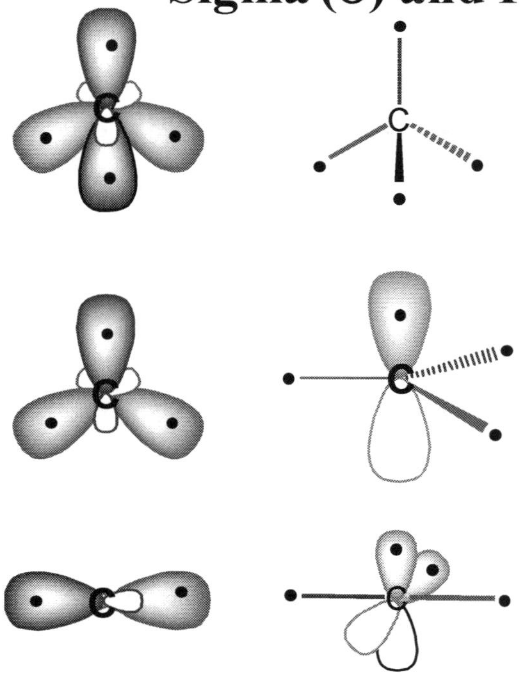


**Here are the pictures we saw for hybrid orbital sets on carbon (left) and the geometry of hybrid orbitals along with leftover p orbitals (shown in grey) on the right. Recall (look at a periodic table) that a carbon atom has four valence electrons, so I have placed one electron in each valence orbital. That means that another atom must supply one electron to form a bond with any of these orbitals ...**

We will consider how each type of hybridized carbon shown will form sigma bonds using the hybrid orbitals and pi bond with leftover p orbitals

**Notes:**

# Sigma (σ) and Pi (π) Bonds

**We can bind two $sp^3$ hybridized carbons together and make sigma bonds to hydrogen atoms with the remaining $sp^3$ orbitals**

**This is ETHANE. There are no more valence orbitals, so we can make no more bonds.**

**Notes:**

# Sigma (σ) and Pi (π) Bonds

**After using all the $sp^2$ hybrids to make sigma bonds, we have a pi orbital left over on each carbon, each with one electron. These two orbitals are close together and have the right number of electrons (two) to form a bond. Not surprisingly, this is what happens, and this is an example of a pi bond!**

Here is a cartoon of what the pi bond looks like. Note that the electrons are above and below the carbon-carbon line.
This molecule is ETHENE.

**Notes:**

# Sigma (σ) and Pi (π) Bonds

**After using all the sp hybrids to make sigma bonds, we have two pi orbitals left over on each carbon, each with one electron. These orbitals are close together and form two pi bonds, so overall there is a triple bond between the carbons!**

Here is a cartoon of what the pi bonds looks like. Note that the electrons are above and below the carbon-carbon line.
This molecule is ETHYNE, commonly called ACETYLENE, a flammable gas used in welding.

**Notes:**

For our purposes:

1. All single bonds are
2. All double bonds are made of
3. All triple bonds are made of

Finally, a note on electronegativity of hybridized carbons:
EN of carbons increases in the order

The s orbital is closer to the nucleus than the p orbital. It is consequently filled with electrons before the p orbital, and we may think of this as demonstrating that an orbital that has more "                    " has a greater pull on electrons.

Since sp is "half s and half p", and $sp^3$ is "$^1/_4$ s and ¾ p", it is logical that the sp hybrid has a greater pull for electrons

**Notes:**

## Bond Angles

Using your knowledge of hybridization and Lewis Structures, predict the following bond angles:

Angles around C in HC(O)H

H–C–C angle in HCCH

H–N–C angle in $H_2NCH_3$

Angles about Al in $AlCl_3$

Angles about B in $BF_4^-$

**Notes:**

# Acid-Base Theory

In General Chemistry we learned about acids and bases, pH, $pK_a$, etc. We will briefly review some of these ideas. There are various ways to think about acid-base reactions. is to use the Brønsted-Lowry definition:

Arrhenius Definition

acid:

base:

Brønsted-Lowry Definition

acid:

base:

Acid:

Base:

Base:

Lewis Definition

acid:

base:

*recall that in water $H^+$ forms the hydronium ion, $H_3O^+$

**Notes:**

## Lewis Acids and Bases

Identify each as a Lewis Acid, Lewis Base, neither, or both. Show the species acting in each role you have assigned to it. Note that the formal charges are shown, but not all lone pairs are shown.

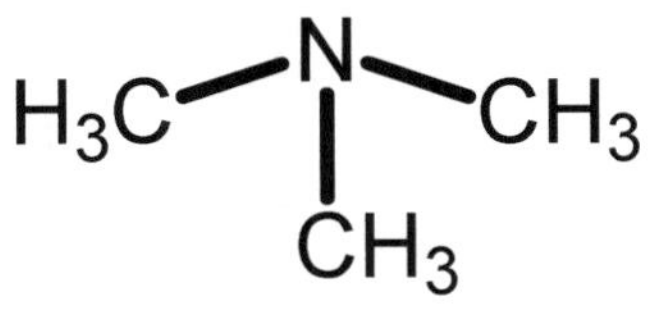


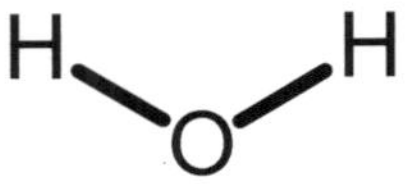


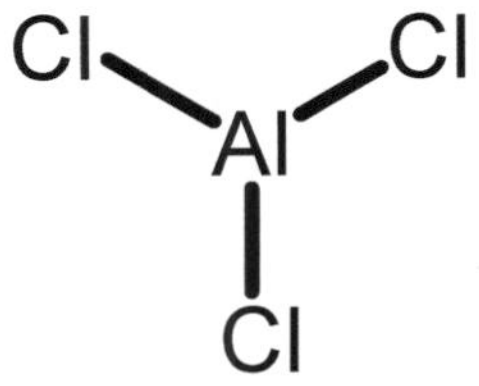


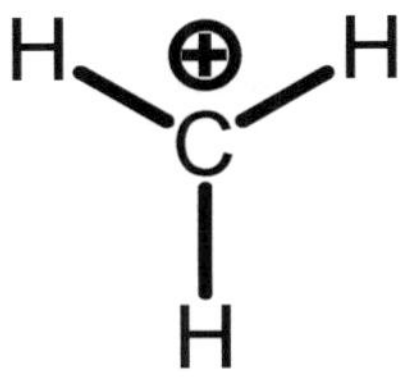


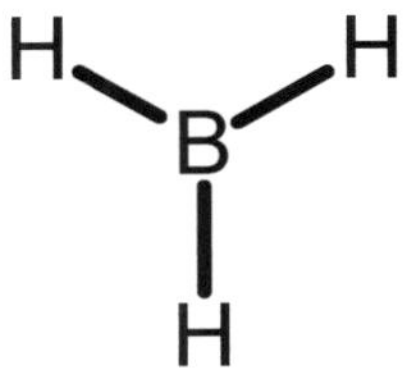


**Notes:**

# Predicting Protonation States.

The 'protonation state' of a chemical species refers to which protons are on the molecule and which are taken off of the molecule. For example, one can envision adding or removing a proton from ammonia to give two new protonation states:

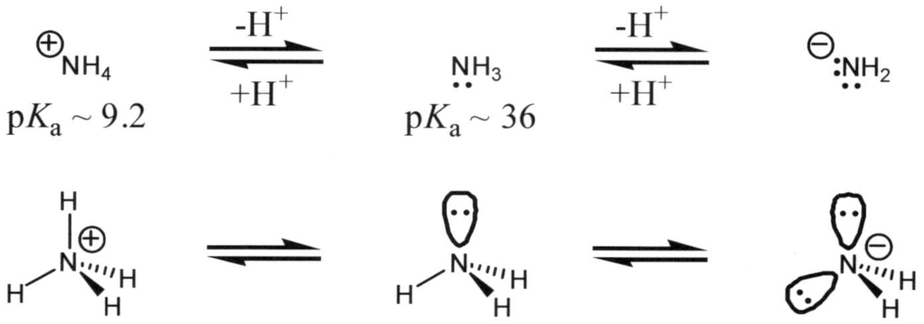


The p$K_a$ can be thought of as **the minimum pH of solution needed to remove the proton**. The p$K_a$ values for ammonia thus allow us to determine the predominant protonation state at pH 7 and at pH 10:

| pH 7: | pH 10: |
| --- | --- |
| | |

**Notes:**

Ⓐ What is the predominant form of acetic acid in a dilute aqueous solution buffered at pH 3?

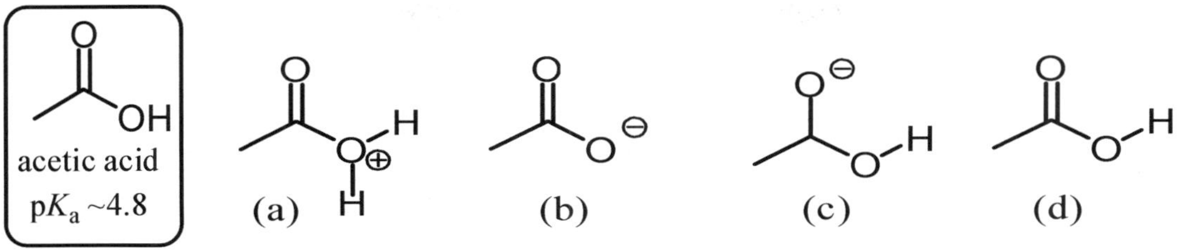


Ⓑ Phosphate is an important biological molecule. What is the predominant protonation state of phosphate in blood (an aqueous solution with pH ~ 7.4)?

| $H_3PO_4$ |
|---|
| $pK_1 \sim 2$ |
| $pK_2 \sim 7.2$ |
| $pK_3 \sim 12$ |

**Notes:**

# Predicting Acidity

$$HA + H_2O \rightleftharpoons A^- + H^+$$

**Any factors that favor dissociation of HA into $H^+$ (to form $H_3O^+$ in water) and $A^-$ will enhance acidity.**

**Factors that affect dissociation:**

***1. ______________________**

**Notes:**

# Effect 1: Anion Size

Within a group (column), the size of the anion has a strong effect on acidity, because:

| | | $pK_a$ ~ |
|---|---|---|
| ________ | HF | ______ |
| | HCl | ______ |
| | HBr | ______ |
| ________ | HI | ______ |

- ORBITAL OVERLAP
- **ANION STABILITY**

**Notes:**

# (cont'd)

**Notes:**

# Effect 2: Electronegativity (EN)

If the atoms to be deprotonated are in the same period (row) then they are about the same size, so the size effects are minimal. Thus, the electronegativity plays a strong role because:

| $CH_4$ | $NH_3$ | $H_2O$ | HF |
|---|---|---|---|

______ ______

$pK_a$ ~ ______ ______ ______ ______

**Notes:**

# Hybridization Influences Electronegativity

If the atom to be deprotonated is the same and only hybridization changes, the hybridization electronegativity plays a strong role because:

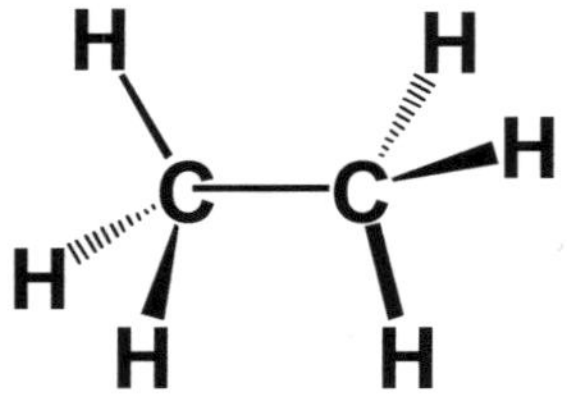
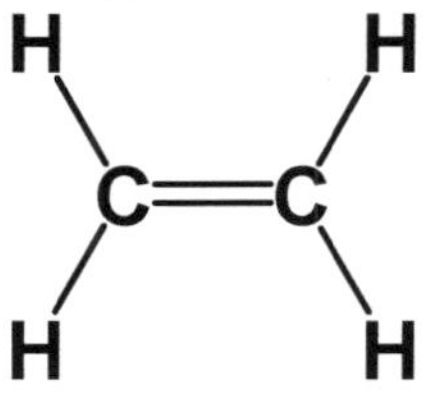

$pK_a$ ~ ________ ________ ________

---

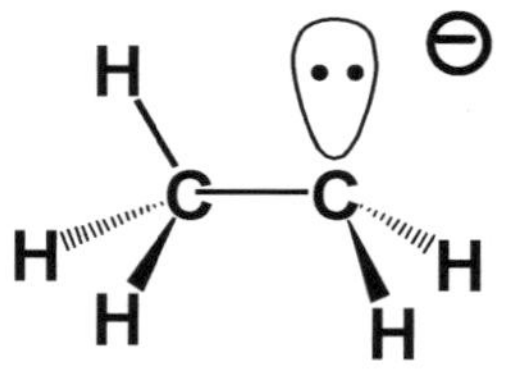
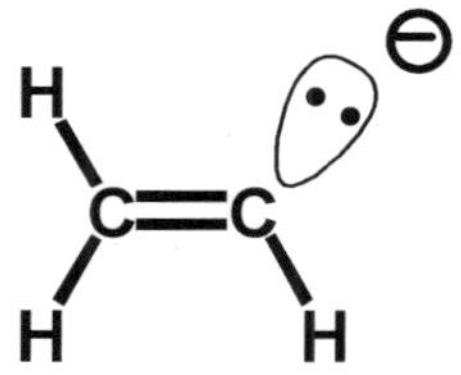
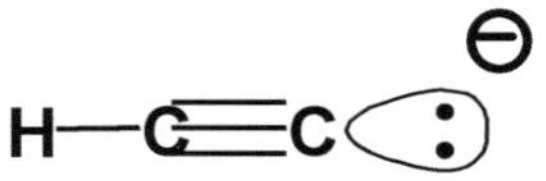

**Notes:**

# Effect 3: "Resonance Effects"

If the anion produced by deprotonation has more than one (good) resonance form, then:

**$CH_3CO_2H$**
$pK_a = 4.7$

**$CH_3OH$**
$pK_a = 15.5$

**HCN**
$pK_a = 9.1$

**Notes:**

# Effect 4: "Inductive Effects"

If an atom to be deprotonated has a partial positive charge INDUCED on it by nearby atoms, it is easier to deprotonate because:

This series illustrates:

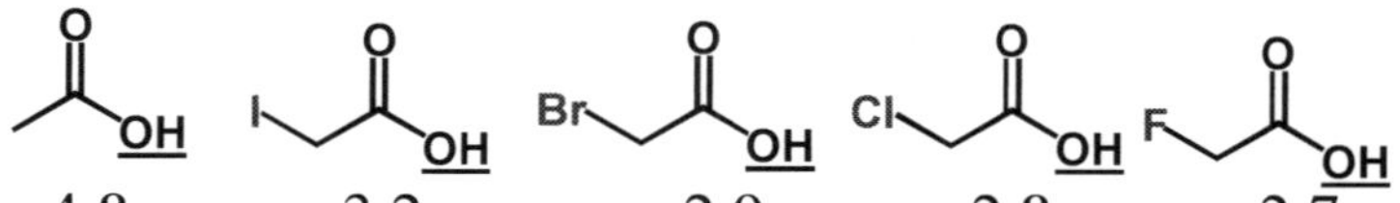


p$K_a$ 4.8 3.2 2.9 2.8 2.7

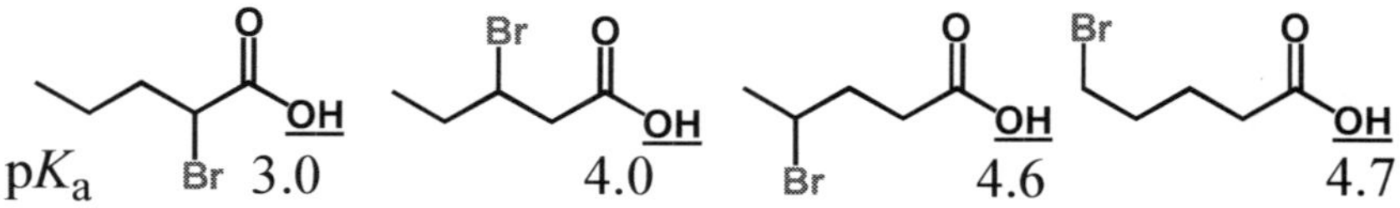


p$K_a$ 3.0 4.0 4.6 4.7

This series illustrates:

**Notes:**

# (cont'd)

(Hyperconjugation)/sterics:

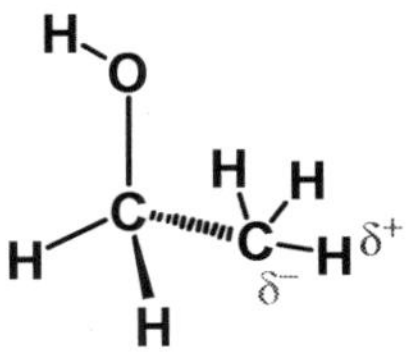

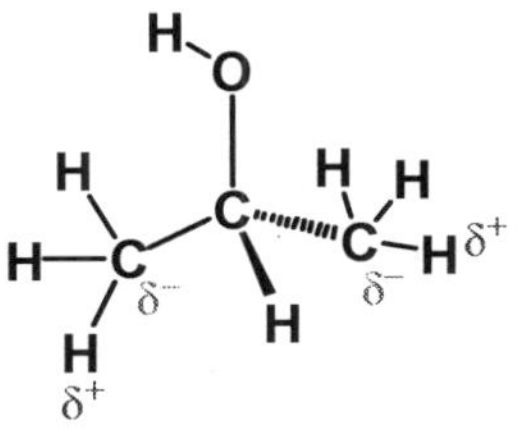

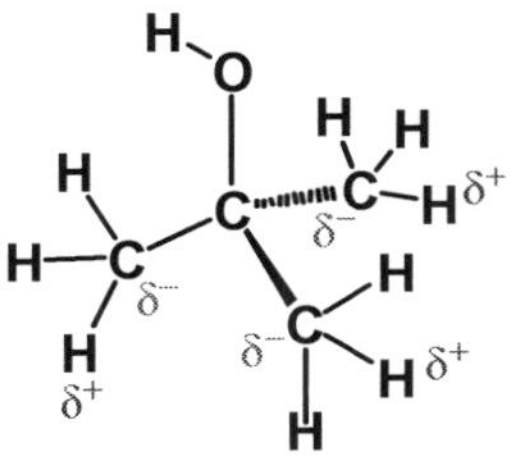

p$K_a$ ~ ______ ______ ______

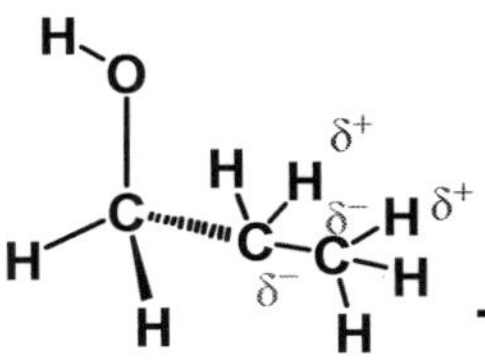

______

**Notes:**

# Acid-Base Example Problems I

Which is the strongest acid:

a) $SiH_4$ b) $PH_3$ c) $SH_2$ d) HCl

Which is the strongest base:

a) $H_3C^-$ b) $H_2N^-$ c) $HO^-$ d) $F^-$

Which has the lowest p$K_a$:

| | | | |
|---|---|---|---|
| $CH_4$ | $H_3C{-}CH_3$ | $H_2C{=}CH_2$ | $HC{\equiv}CH$ |
| (a) | (b) | (c) | (d) |

**Notes:**

# Acid-Base Example Problems II

Which proton shown in each hypothetical molecule is most acidic?

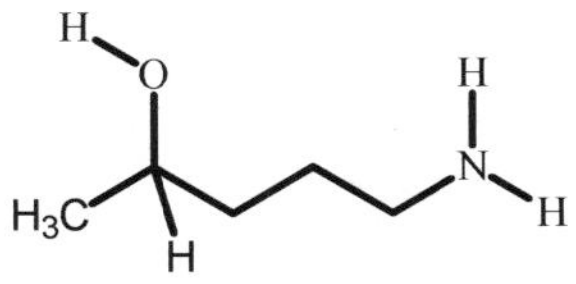

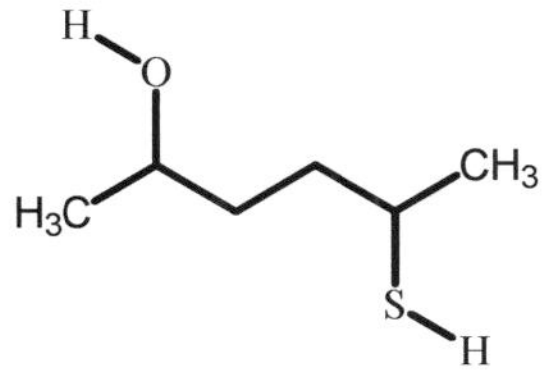

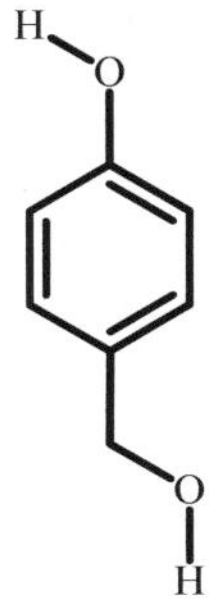

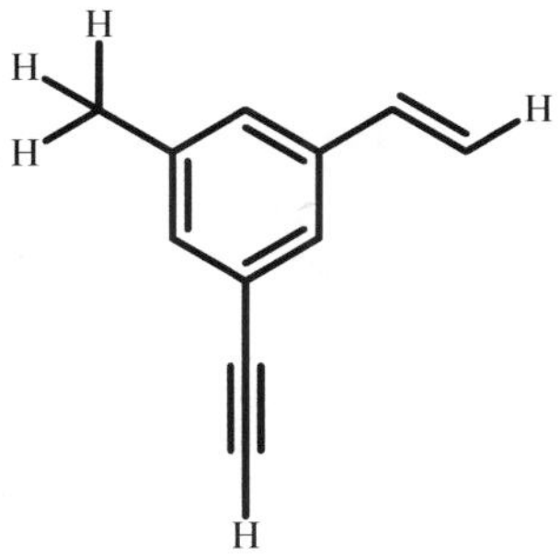

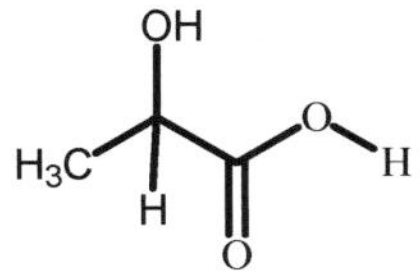

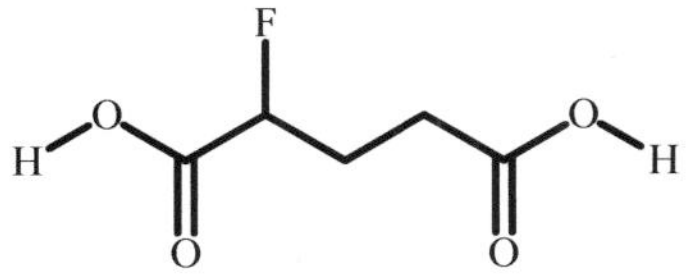

**Notes:**

# Acid-Base Reaction Predictions

Reactions generally proceed spontaneously to produce more stable anions (weaker bases). What would you predict to be the products of these reactions:

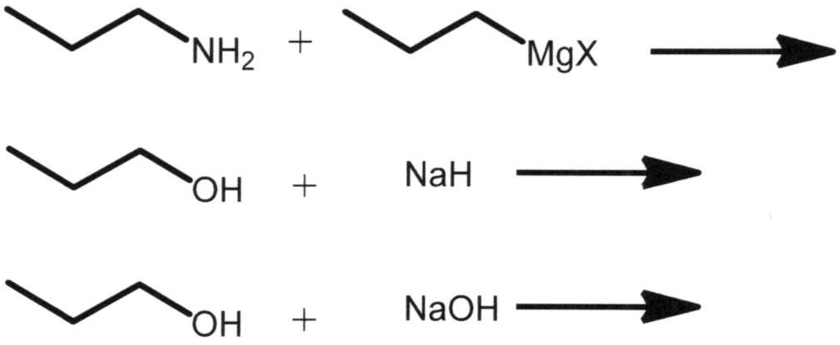


**Notes:**

# Relative Stability of Cations

Ⓐ **Observation:**

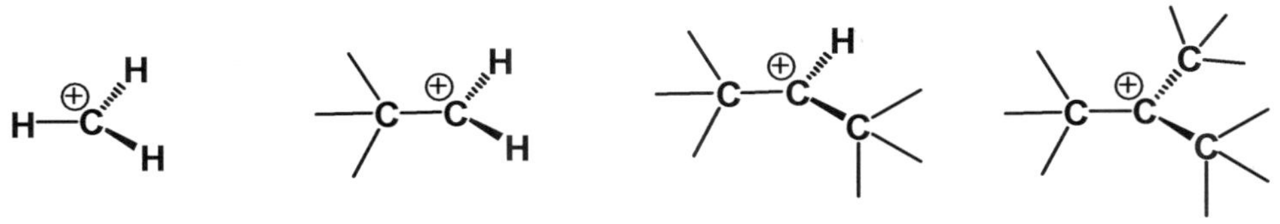


**methyl** **1°** **2°** **3°**

**STABILITY**

**LESS STABLE** **MORE STABLE**

Ⓑ **Explanation:**

**Notes:**

# Resonance and Carbocations

Increasing
stability of cations

**Notes:**

# Relative Stability of Radicals

Increasing stability
of radicals

**Notes:**

# Resonance and Stability

When given a structure and tasked with determining its relative stability, one must consider all resonance contributors to the real structure.

Which in each set is the most stable species?

**Set 1:**

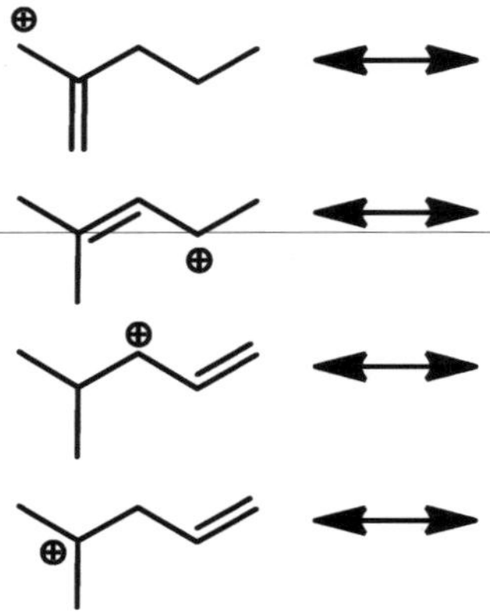

**Set 2:**

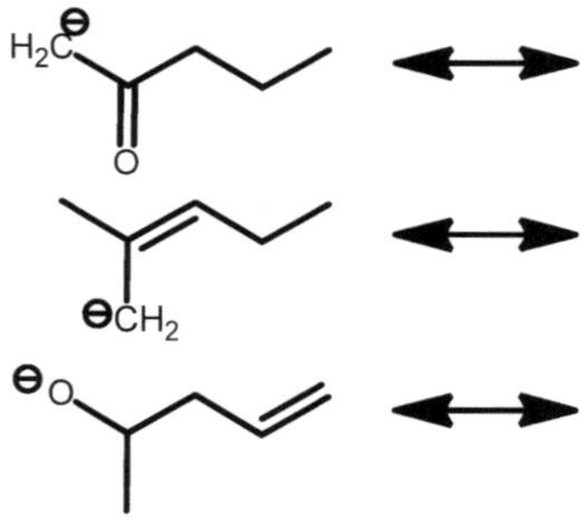


**Notes:**

# Lecture Set 2: Nomenclature and Functional Groups

**Suggested Reading:**

**Suggested Problems:**

# Functional Groups I

## Hydrocarbons

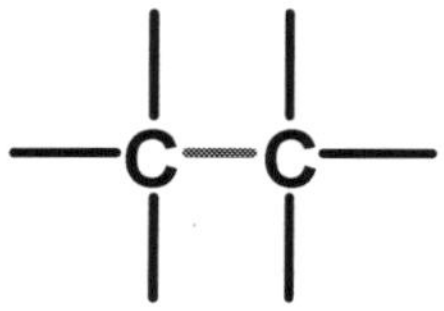

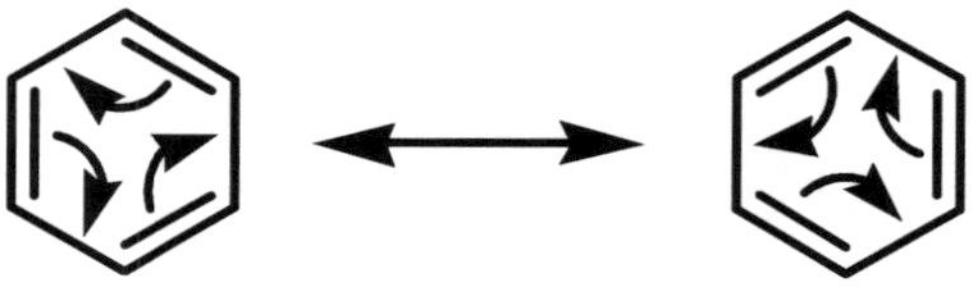

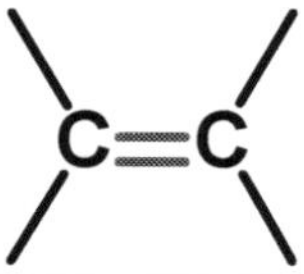

**Notes:**

# Functional Groups II

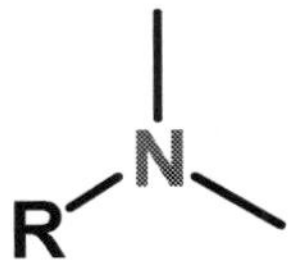

R—OH

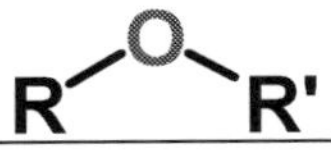

R—X

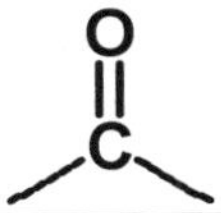

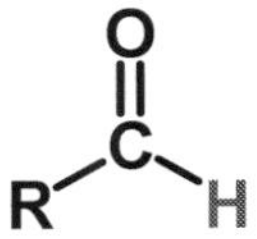

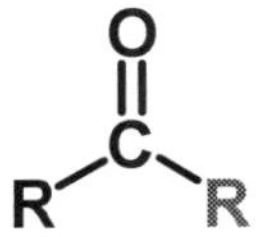

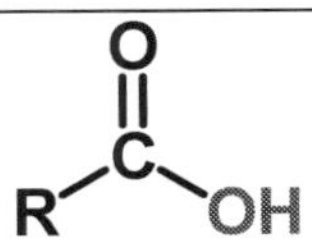

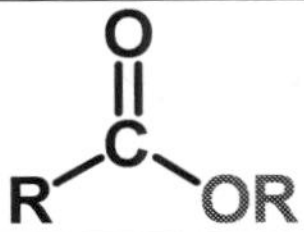

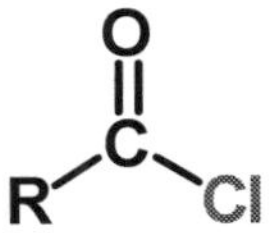

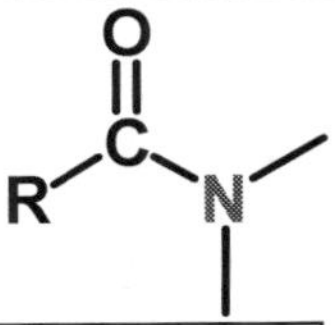

**Notes:**

# Organic Chemistry: Functional Groups

**Organic chemistry:** The chemistry of **functional groups**- groups of atoms which behave similarly in a variety of molecules.
Your first step is to learn the names of the functional groups (FGs)

Example. In the molecules below, circle all the functional groups you can identify. Write the name of the functional groups. Identify each compound as hydrocarbon (HC) or heteroatom-containing (HA).

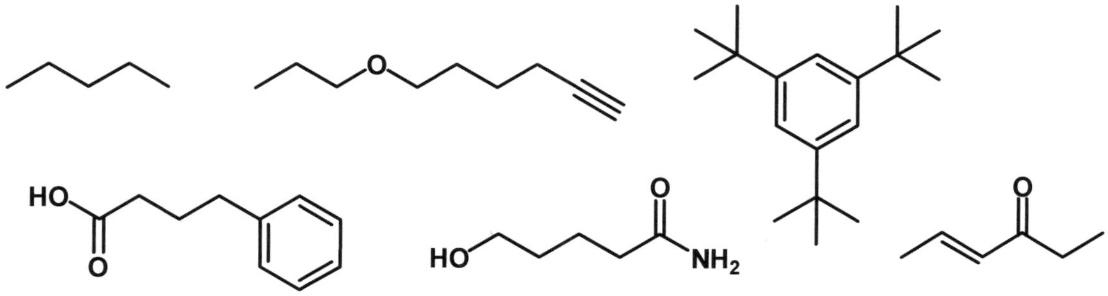


**Taxol:**

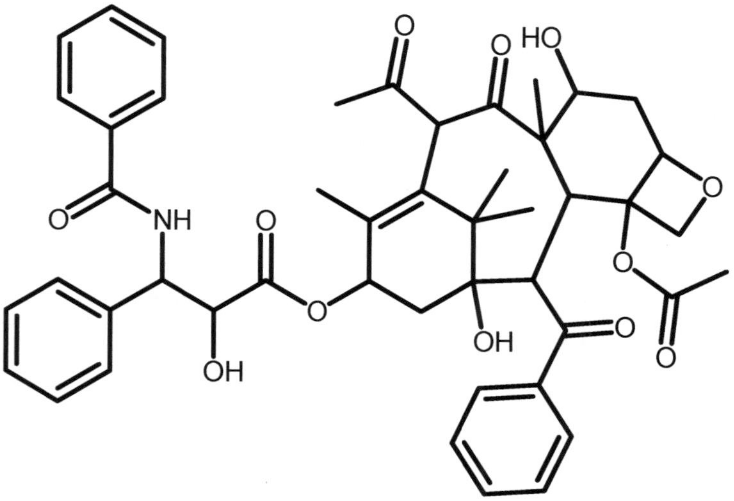

# Nomenclature: Linear Alkanes

| | | |
|---|---|---|
| $CH_4$ | | |
| $CH_3CH_3$ | | |
| $CH_3CH_2CH_3$ | | 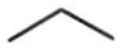 |
| $CH_3(CH_2)_2CH_3$ | |  |
| $CH_3(CH_2)_3CH_3$ | |  |
| $CH_3(CH_2)_4CH_3$ | |  |
| *n*-$C_7H_{16}$ | | 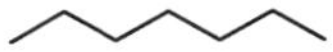 |
| *n*-$C_8H_{18}$ | |  |
| *n*-$C_9H_{20}$ | |  |
| *n*-$C_{10}H_{22}$ | | 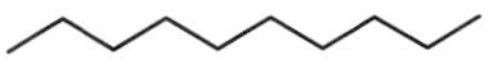 |

**Notes:**

# Constitutional Isomers of Alkanes: $C_4H_{10}$

Constitutional Isomers-

$C_4H_{10}$

linear alkanes
('*n*ormal' alkanes)

$CH_3-CH_2-CH_2-CH_3$

branched alkanes

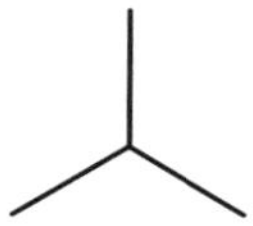

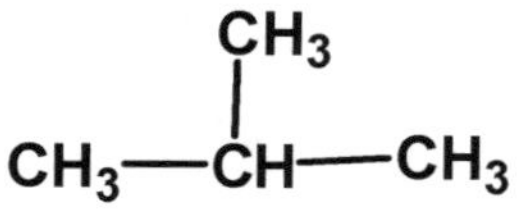

Lighter fluid is 5% *n*-butane and 95% isobutane

**Notes:**

# Constitutional Isomers of Alkanes: $C_5H_{12}$

## $C_5H_{12}$

$CH_3-CH_2-CH_2-CH_2-CH_3$

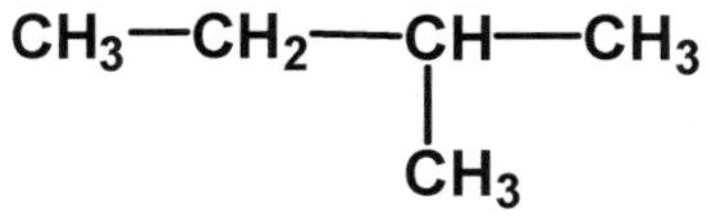


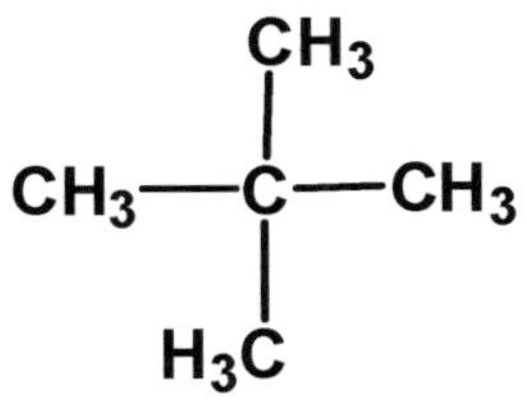


**Notes:**

# Alkane Problem

Example. Draw all the constitutional isomers possible for these formulae:

$C_4H_{10}$(2)

$C_5H_{12}$ (3)

$C_6H_{14}$ (?)

$C_7H_{16}$ (9)

↑

**# of constitutional isomers**

**Notes:**

# Properties: Branched vs. Linear Alkanes

**Structure-property relationships:**

**Obviously the structure of a molecule has an impact on the properties of that molecule, otherwise all molecules would have the same properties.**

**In our effort to rationalize the properties of alkanes, we will focus on:**

**1. The ____________ of the alkane**
**2. Whether the alkane is ____________**

**We will only look at a couple properties in this lecture:**

**1. ____________**

**2. ____________**

**Why do we study only these???**
**The heat of combustion tells us about ____________**
**The boiling point tells us about ____________**

**Notes:**

# Properties: Branched vs. Linear Alkanes

## Explosives (high energy materials) are unstable- that's why they explode.

Heats of combustion ($\Delta H_c$) for burning $C_8H_{18}$ isomers:

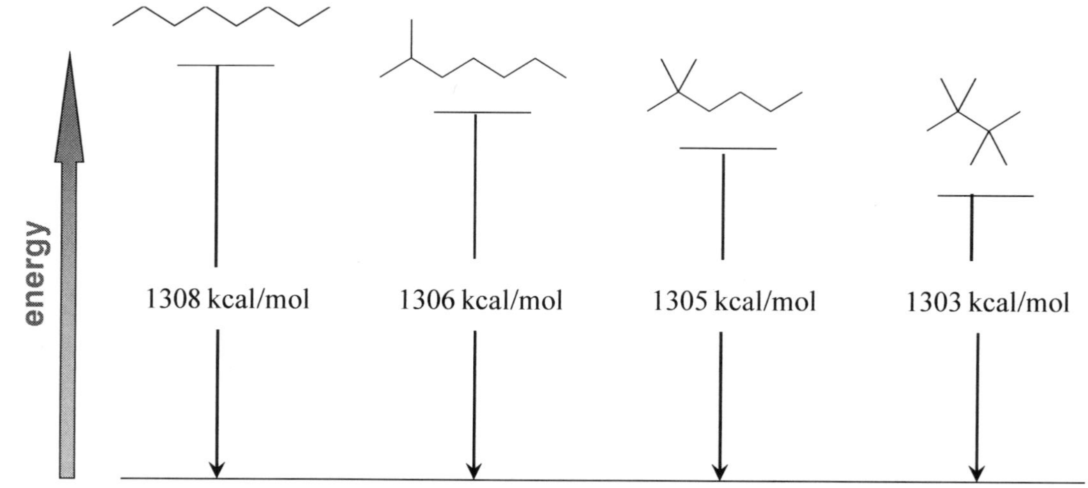


The **more branching**, the **less heat released**, so
**more branched =**

**Notes:**

# Properties of Alkanes

General Chemistry: Essentially nonpolar molecules like alkanes have intramolecular forces called
These are relatively weak intermolecualr forces compared to other types, but **the more interaction present, the more strongly the molecules are held together.**

Easier to boil

| alkane | bp |
|---|---|
| $CH_4$ | -160°C |
| $C_2H_6$ | -89 |
| $C_3H_8$ | -42 |
| *n*-$C_4H_{10}$ | -0.4 |
| *n*-$C_5H_{10}$ | 36 |

Compare linear to branched alkanes:

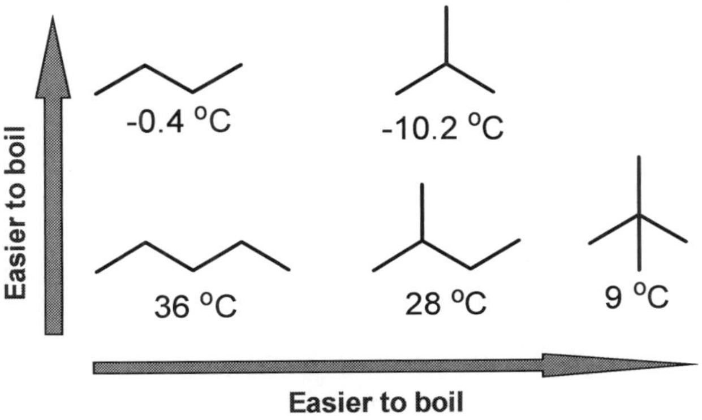


The **lighter** or **more branching**, the **less intermolecular force**, so
**more branched =**
**Lighter =**

**Notes:**

# Carbon Types

## A

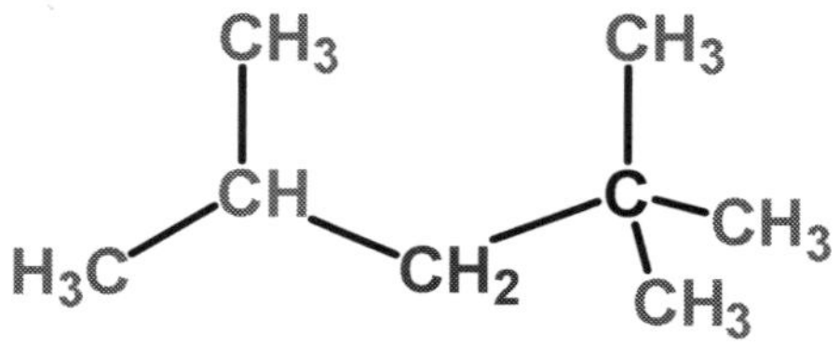


$CH_3$ =

$CH_2$ =

CH =

C =

## B

$CH_4$ = methane

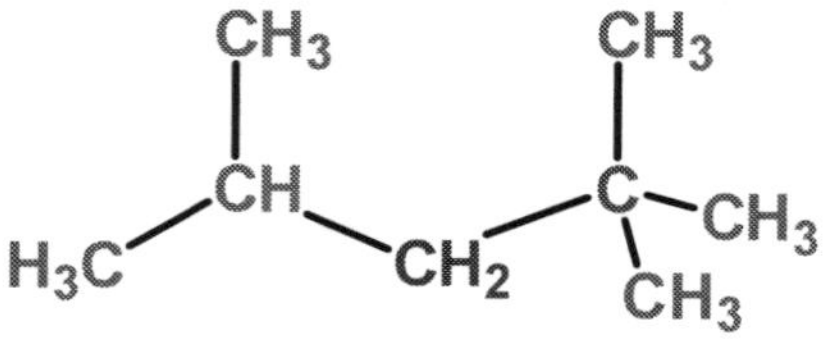


$CH_3$ =

$CH_2$ =

CH =

**Notes:**

# Substituents

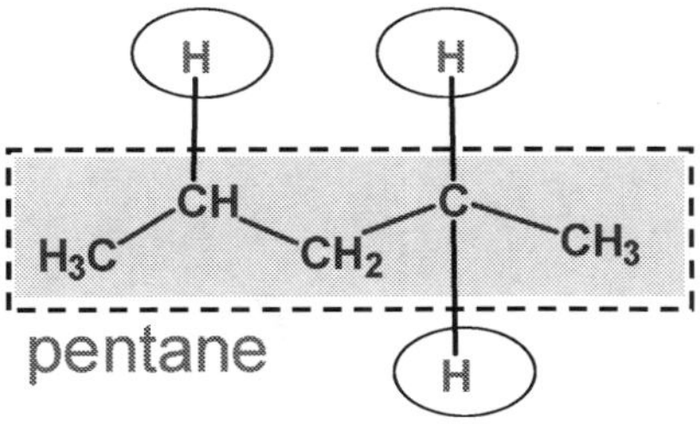


pentane

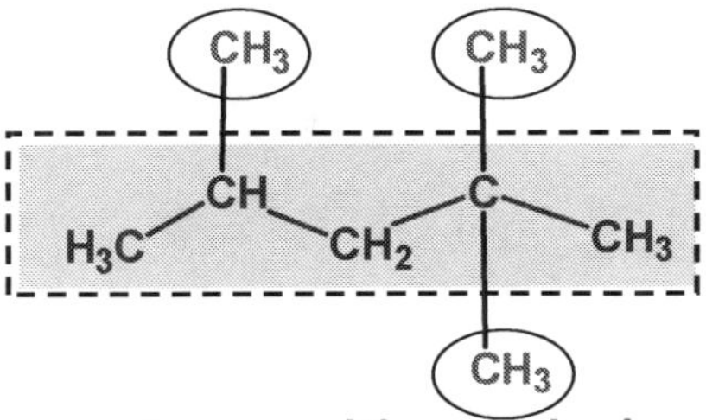


pentane with methyl groups
substituted for some of the hydrogens

| Substituent: |
| --- |
| |

Substituents that occur frequently in molecules of interest are given names to make naming a complex molecule more achievable.

Of course there are substituents other than methyl groups we can put on to a molecule …

**Notes:**

# Substituent Names: *n*-Alkyl Groups

| | | | |
|---|---|---|---|
| methane | $CH_4$ | —$CH_3$ | methyl |
| ethane | 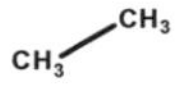 | 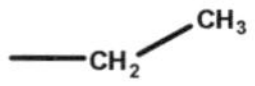 | ethyl |
| propane | 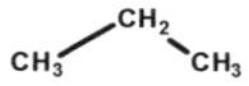 | 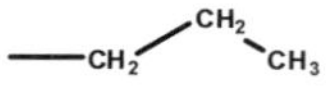 | n-propyl |
| n-butane | 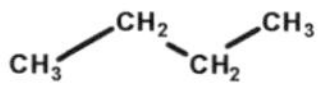 | 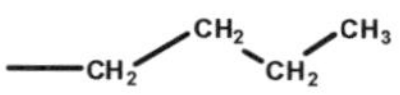 | n-butyl |
| n-pentane | 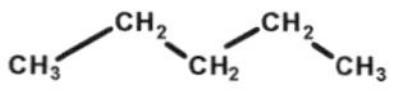 | 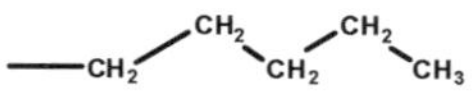 | n-pentyl |
| n-hexane | 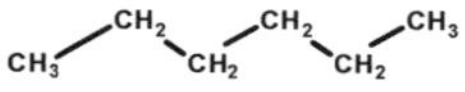 | 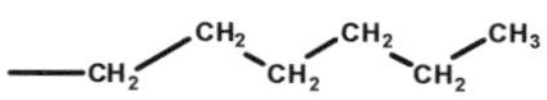 | n-hexyl |
| n-heptane | 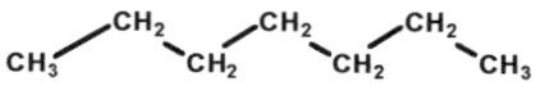 | 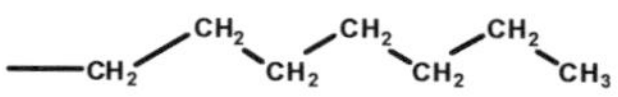 | n-heptyl |
| n-octane | 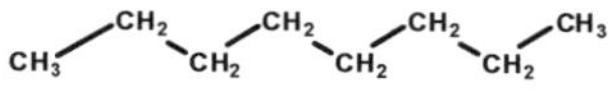 | 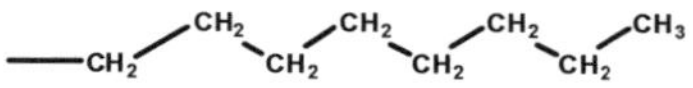 | n-octyl |
| n-nonane | 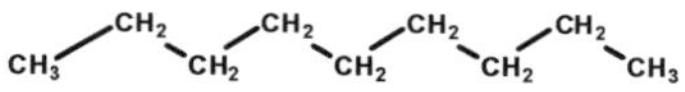 | 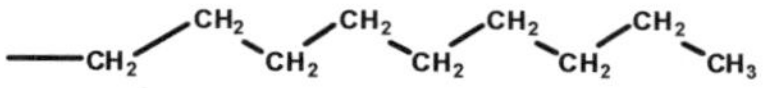 | n-nonyl |
| n-decane | 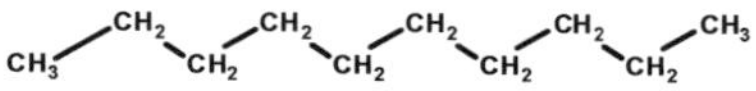 | 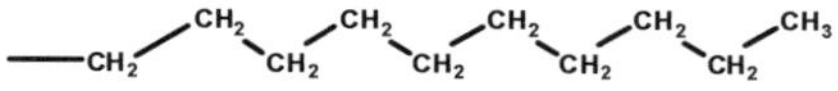 | n-decyl |

**Notes:**

# Substituent Names

If you substitute the "end H", just [ ]

**propane**

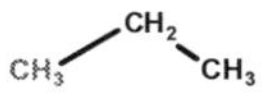


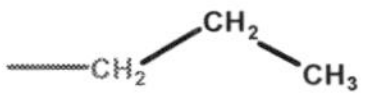


***n*-propyl** "*n-*" indicates it is "normal", linear

**propane**

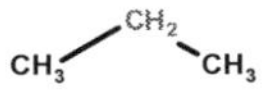


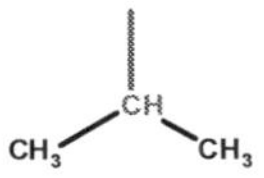


**isopropyl** "*iso-*" indicates it is branched

***n*-butane**

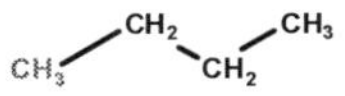


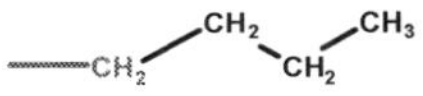


***n*-butyl** "*n-*" indicates it is "normal", linear

***n*-butane**

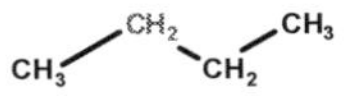


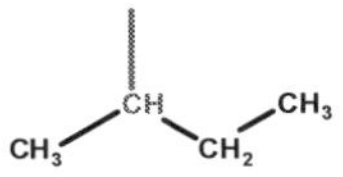


***sec*-butyl** "*sec-*" indicates it is substituted at the secondary carbon

**isobutane**

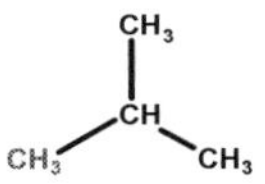


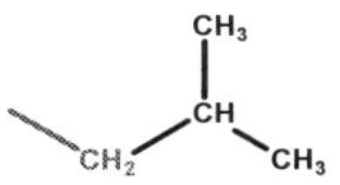


**isobutyl** if you substitute the end H, just change "-ane" to "-yl"

**isobutane**

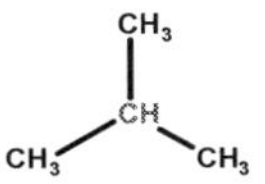


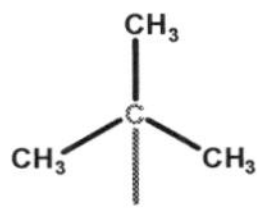


***tert*-butyl** "*tert-*" indicates substitution at the tertiary carbon

**Notes:**

# Nomenclature: Branched Alkanes

## Your step-by-step guide

1.

This is the 'parent chain'; the other stuff on the parent chain will be named as substituents. Texts and profs will draw the molecule in weird ways to try to fool you, so look out:

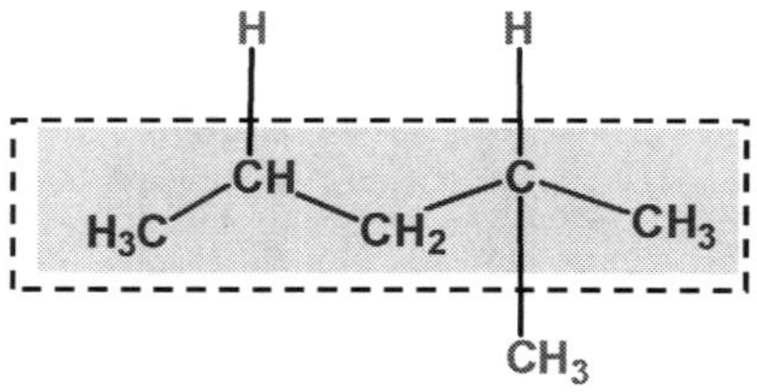


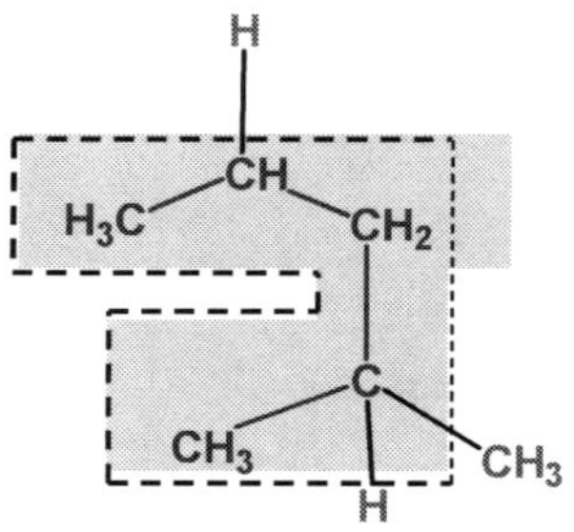


**Notes:**

# Nomenclature: Branched Alkanes

2. Number the carbon atoms in the 'parent chain' in the way that gives the lowest number to the substituent closest to an end of the parent chain.

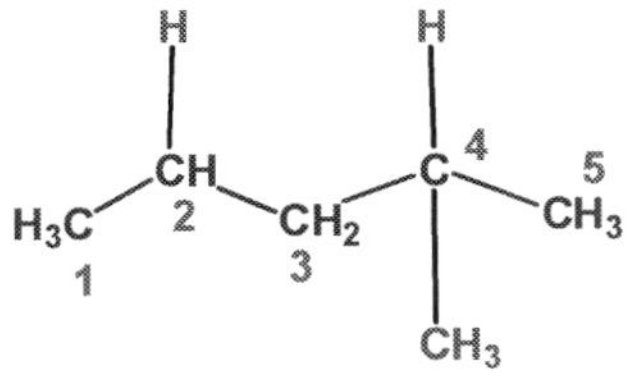


or

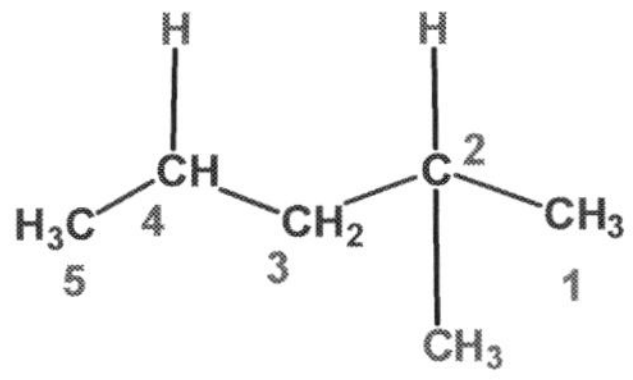


?

(A)

(B)

(C)

This molecule is properly named

**Notes:**

# Nomenclature

Here is an alkane; find longest chain

**alkane** **possible parent chains:**

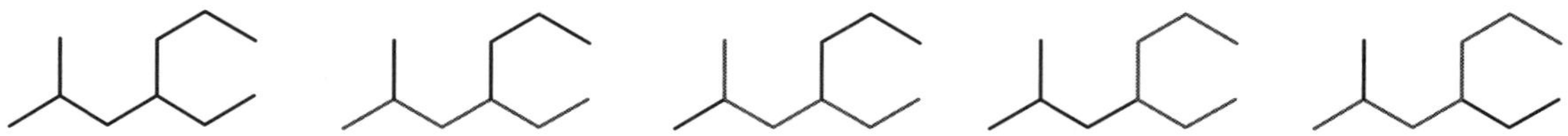

**#C's: 6 6 6 7**

Number the parent chain to give lowest number to substituent closest to end

**possible numbering schemes:**

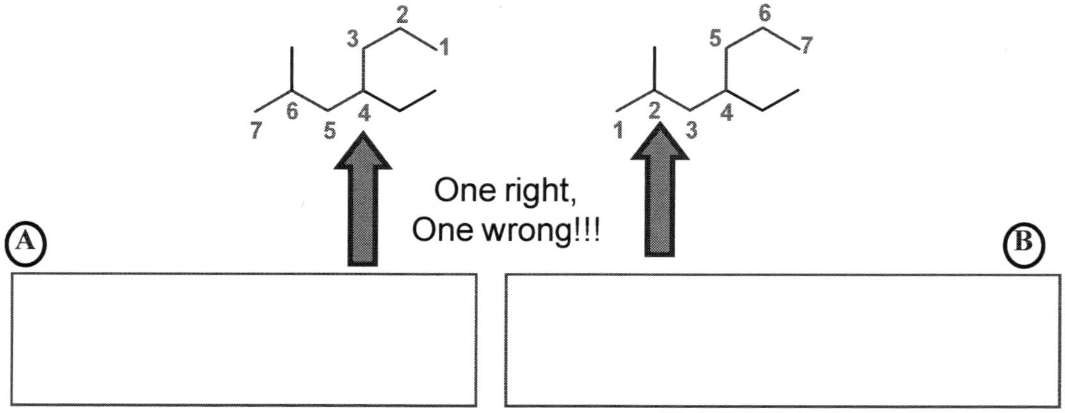


**Notes:**

# Alphabetize

3. With more than one type of substituent,

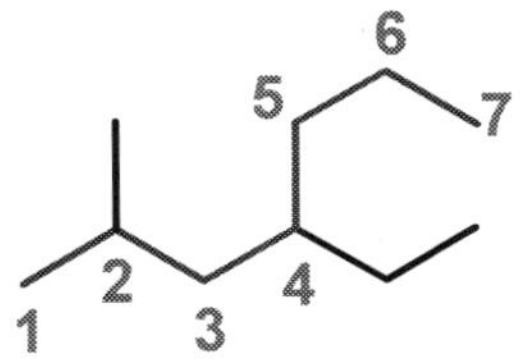


We could name it
2-methyl-4-ethylheptane
or
4-ethyl-2-methylheptane

Since 'ethyl' is alphabetically before 'methyl', this molecule is properly named

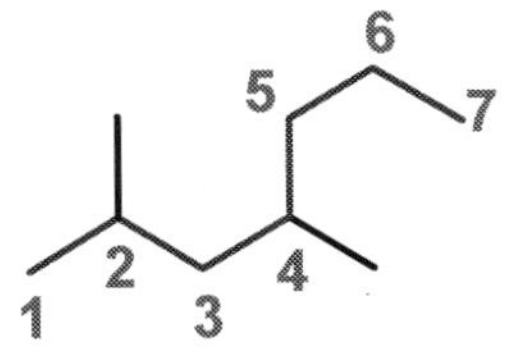


Now consider this closely related molecule, in which I changed the ethyl group to a methyl group. The numbering and parent chain are the same, but now we have two methyl groups; we need to apply the next rule ...

**Notes:**

# Prefixes

4. If more than one of the same substituent are present on your parent chain, use di, tri, tetra, etc. prefixes to denote this (these prefixes do not count when alphabetizing, though; neither do the n-, sec-, or tert- prefixes)

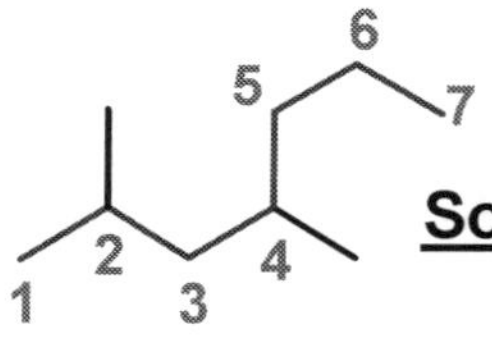


**So, we call this**

NOT
~~2-methyl-4-methylheptane~~
or
~~4-methyl-2-methylheptane~~

What **NOT** to write on the exam!!!!

**Notes:**

# Numbering Ambiguity

There are even more potential complications that could trip us up; consider:

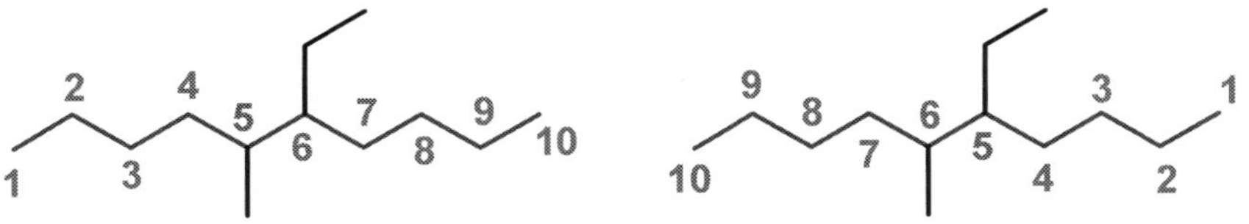


We've easily found the parent chain (decane), and we try to find the right numbering scheme; but either way we count, the first substituent is at the **5** position; which is right!?

We need the next rule:

| |
|---|
| 5. If numbering leads to the same "lowest number" substituent in both directions the correct numbering |

In this case, ethyl is alphabetically before methyl, so ethyl is given the 5:

This molecule is properly named

**Notes:**

# Parent Chain "Ambiguity"

The previous example showed you how to deal with substituent numbering issues; but what if you have trouble identifying the parent chain to begin with? Consider:

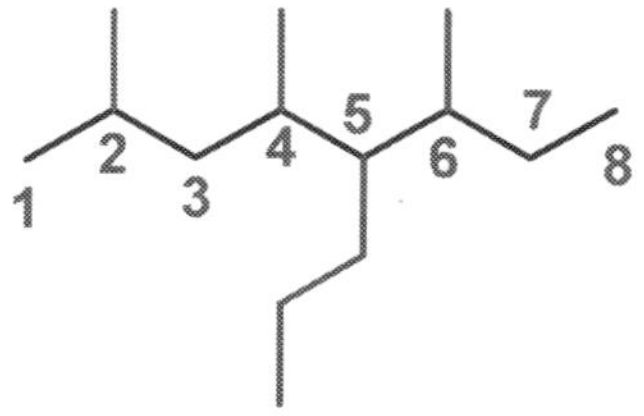


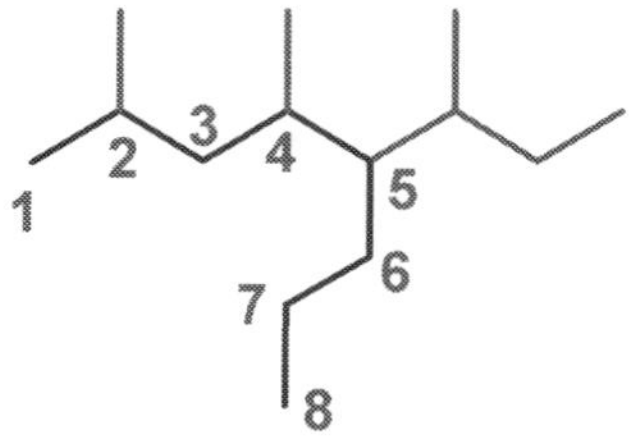


We've found **two** possible octane parent chains; the one on the left with four substituents, and the one on the right with three substituents. Which is right?! We need another rule.

6. If you find two different possible parent chains of the same length,

The one on the left has more substituents, so we use that parent chain.

This molecule is properly named

**Notes:**

# Complicated Substituents Rule

7. Substituents that are not *n*-alkyl groups will need to be named using their own nomenclature rules:

    a. Always give number 1 to the carbon attached to the parent chain
    b. The first letter of the substituent name is used for alphabetizing
    c. In the final name of the whole compound, these 'complicated substituent' names are put in parentheses to cut down on the confusion.

Consider:

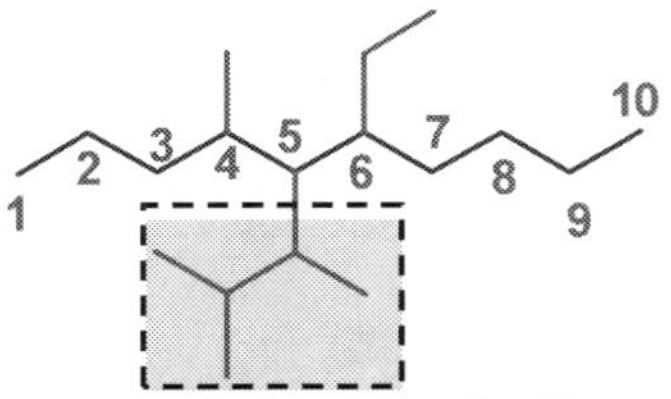


We know the name will be 5-(substituent name)-6-ethyl-4-methyldecane.
Name the sidechain:

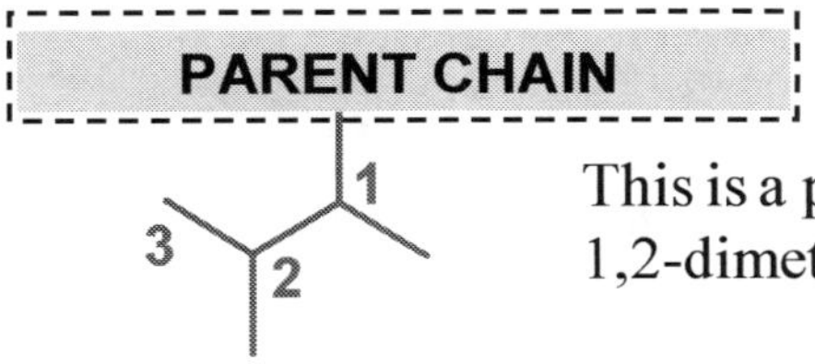


This is a propyl with two methyls on it, so it will be called: 1,2-dimethylpropyl

The molecule is properly named 5-(1,2-dimethylpropyl)-6-ethyl-4-methyldecane

**Notes:**

## Your step-by-step guide to alkane nomenclature

| |
|---|
| 1. Find the longest chain. This is the 'parent chain'; the other stuff on the parent chain will be named as substituents. Texts and profs will draw the molecule in weird ways to try to fool you, so look out: |
| 2. Number the carbon atoms in the 'parent chain' in the way that gives the lowest number to the substituent closest to an end of the parent chain. |
| 3. With more than one type of substituent, name in alphabetical order. |
| 4. If more than one of the same substituent are present on your parent chain, use di, tri, tetra, etc. prefixes to denote this (these prefixes do not count when alphabetizing, though; neither do the n-, sec-, or tert- prefixes) |
| 5. If numbering leads to the same lowest number substituent in both directions the correct numbering gives the lowest number to the substituent that is first alphabetically. |
| 6. If you find two different possible parent chains of the same length, you choose the one with more substituents coming off of it. |
| 7. Substituents that are not *n*-alkyl groups will need to be named using their own nomenclature rules:<br>a. Always give number 1 to the carbon attached to the parent chain<br>b. The first letter of the substituent name is used for alphabetizing<br>c. In the final name of the whole compound, these complicated substituent names are put in parentheses to cut down on the confusion. |

**Notes:**

# Nomenclature Problem

Example. Name these compounds.

1 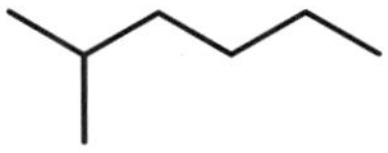

2 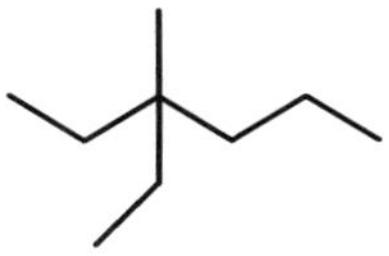

3 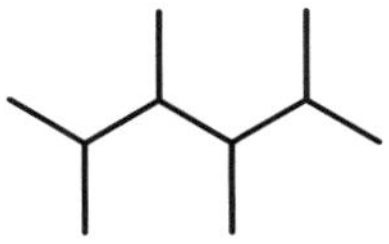

4 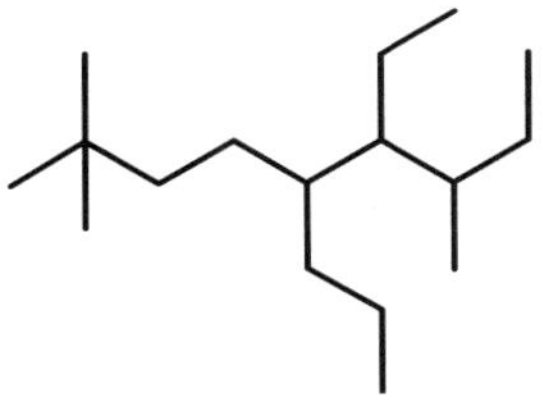

5 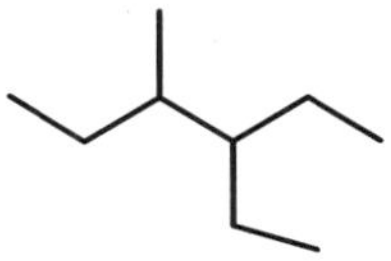

6 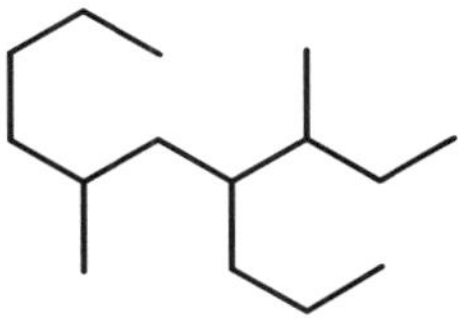

7 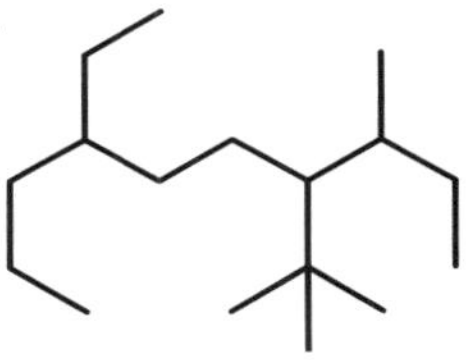

Pointer:
Dashes only go:
1.

2.

**Notes:**

# Nomenclature Answer Part 1

| RULES NEEDED: | STRUCTURE: | NAME: |
|---|---|---|
| 1 and 2 | 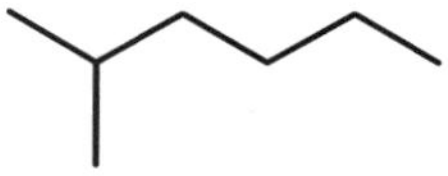 | |
| 1, 2, and 3 | 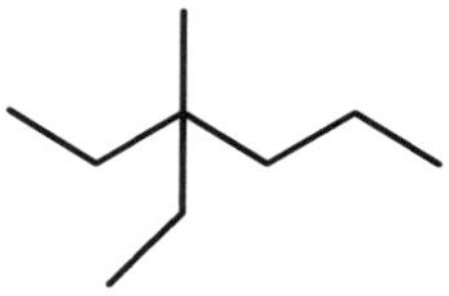 | |
| 1, 2 and 4 | 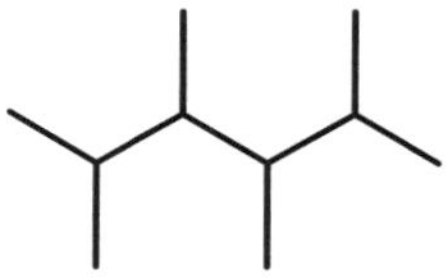 | |
| 1, 2, 3 and 4 | 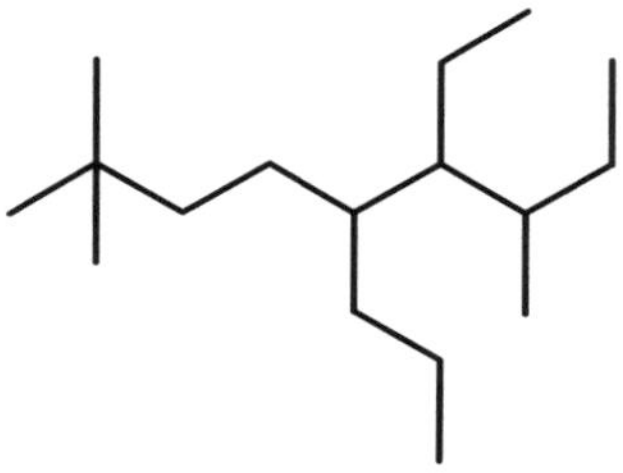 | |

**Notes:**

# Nomenclature Answer Part 2

| **RULES NEEDED:** | **STRUCTURE:** | **NAME:** |
|---|---|---|
| 1, 2, 3 and 5 | 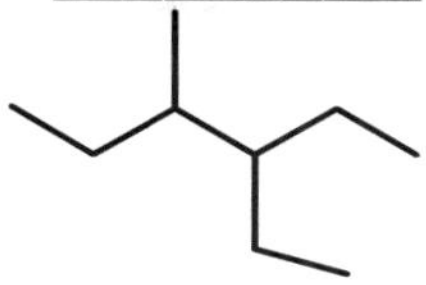 | |
| 1, 2, 3, 4 and 6 | 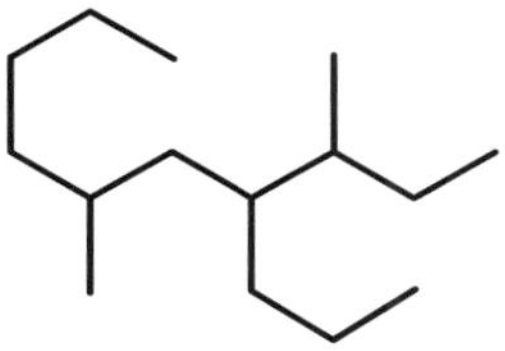 | |
| 1, 2, 3, 4 and 7 | 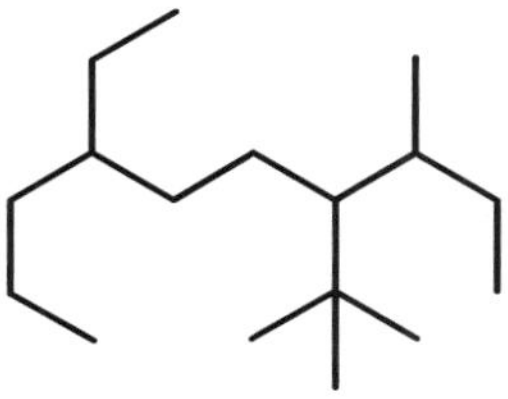 | |

Each subsequent structure requires more rules to correctly solve. **This is the type of nomenclature problem you can expect to see on the exam.** You should also be able to draw a structure given the name (easier!).

## Notes:

# Cycloalkanes

Cyclic versions of alkanes are called cycloalkanes, and are named by the prefix used in the linear alkane with the same number of carbon atoms:

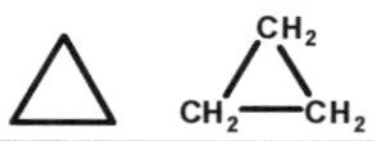

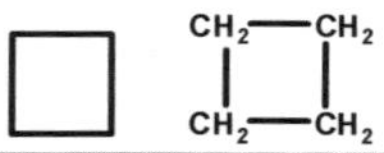

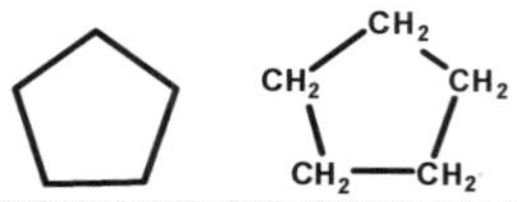

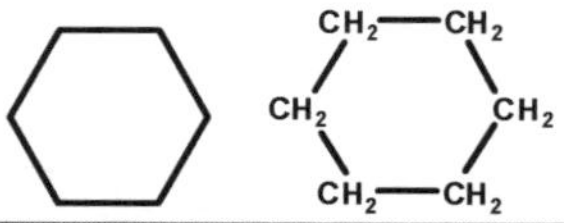

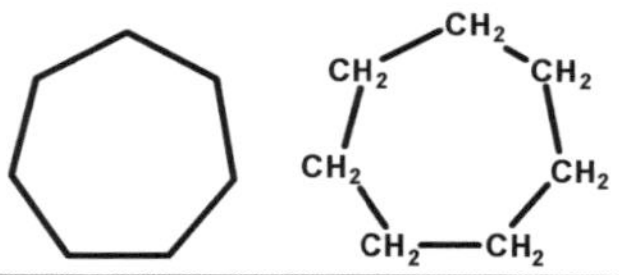

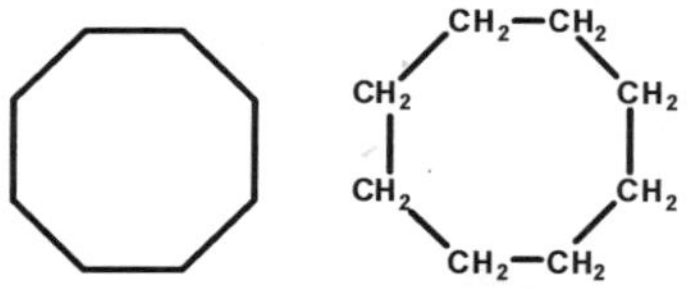

Of course, cycloalkanes can have substituents as well, and we need to know how to name substituted cycloalkanes.

**Notes:**

# Cycloalkanes

Armed with our knowledge of naming branched alkanes, naming cycloalkanes is pretty straightforward. We can use the cycloalkane as the parent chain (if it is longest) and follow numbering/alphabetizing rules from there.

An additional rule for naming cycloalkanes:
If there is a possible parent chain on the cycloalkane that is longer than the # of carbons in the cycloalkane, use that as the parent chain, and name the cycloalkane as a substituent. For example:

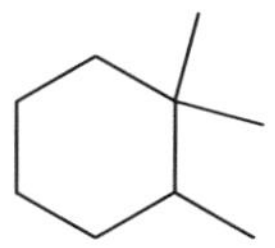

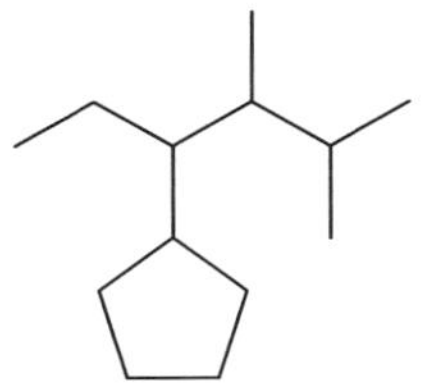

**Notes:**

# Configurational Isomers

Depending on where we put the methyl substituents, we have two different configurational isomers.

A configurational isomer in which substituents are on the same side (both 'up' in this case) is referred to as a *cis-* isomer, while a configurational isomer with substituents on opposite sides is referred to as the ***trans-*** isomer.

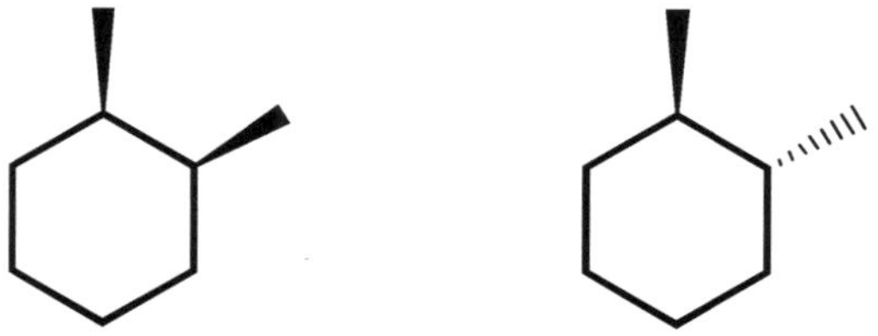

**Notes:**

# Cycloalkanes Problem

Example. Name these compounds.

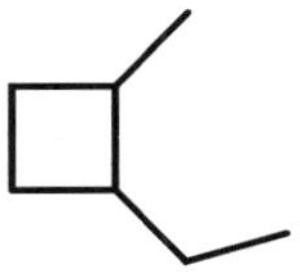

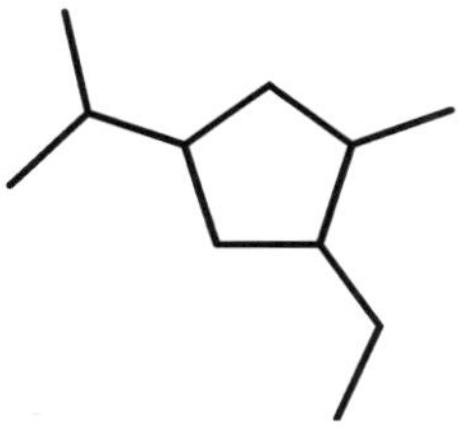

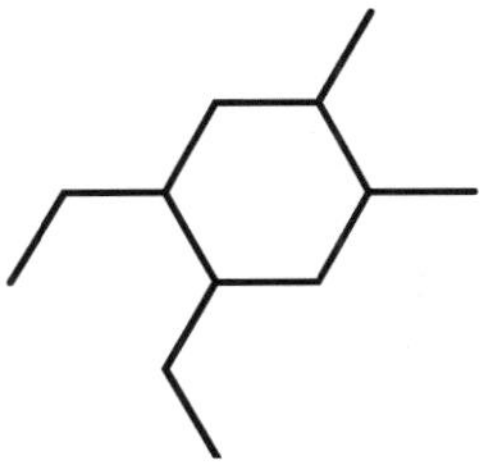

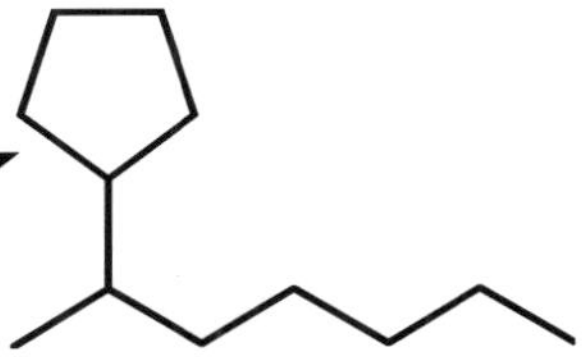

**Notes:**

# Naming Alkyl Halides

For alkyl halides, name the compound using the rules for naming alkanes and cycloalkanes, and treat the halides as substituents:

$CCl_2H_2$ dichloromethane

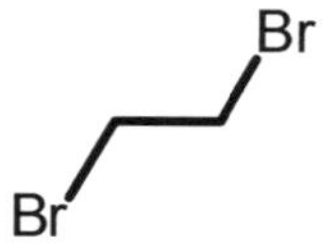


1,2-dibromoethane

-F = fluoro
-Cl = chloro-
-Br = bromo-
-I = iodo-

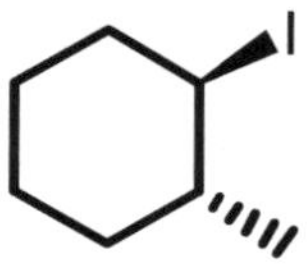


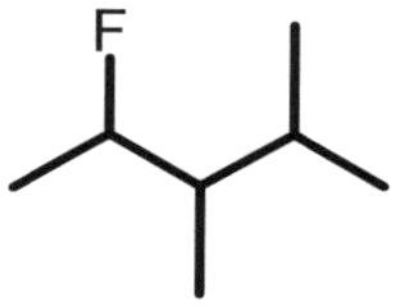


**Notes:**

# Naming Alcohols: Rules

For alcohols, use the rules for naming alkanes and cycloalkanes as a starting point, with the following adjustments:

1. replace the “e” at the end of the alkane name with “ol”.

2. The alcohol is always given the lowest possible number. Note that this means that the alcohol is always given the “1” position in cycloalkanes (so, no need to add a number; we know it’s always 1).

3. Place the number indicating the position of the alcohol directly before the parent chain name (which now ends in “ol”).

**Notes:**

# Naming Alcohols: Examples

$CH_3OH$ Methanol

$CH_3CH_2OH$ Ethanol

1-propanol
(or *n*-propanol)

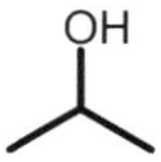


2-propanol
(or isopropanol)

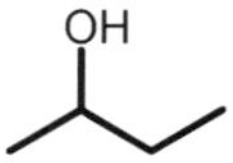


2-butanol
(or *sec*-butanol)

2-methyl-2-propanol
(or *tert*-butanol)

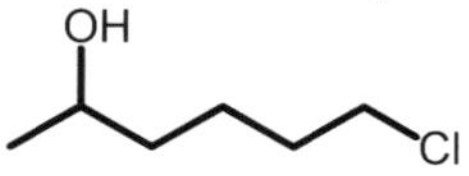


6-chloro-2-hexanol

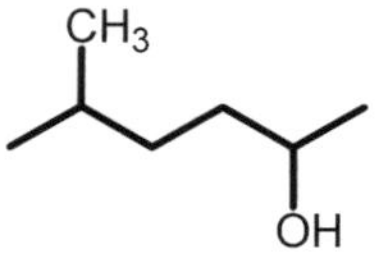


5-methyl-2-hexanol

*trans*-4-methylcyclohexanol

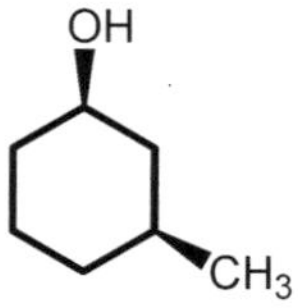


*cis*-3-methyl-cyclohexanol

## Notes:

# Naming Alcohols

Example. Name these compounds.

$FH_2COH$

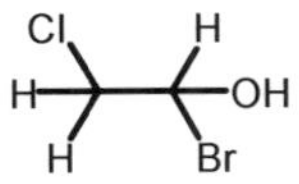

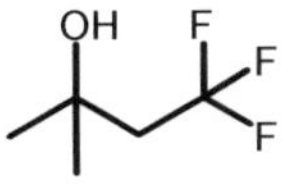

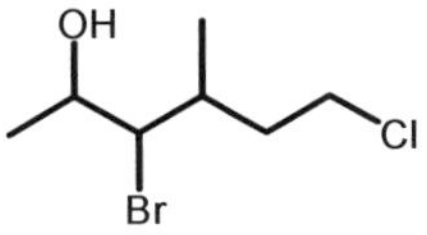

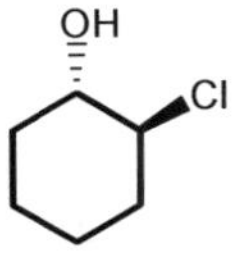

**Notes:**

# Classifying Alcohols and Alkyl Halides

We learned to classify $CH_3$- units as methyl (1°), $-CH_2$- as methylene (2°), etc., we can also classify alcohols similarly. These classifications are very important, because the different classes of alcohols and alkyl halides have different reactivities.

$HO-CH_2R$

= primary alcohol (1° alcohol)

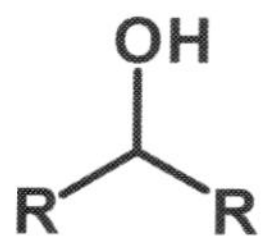


= secondary alcohol (2° alcohol)

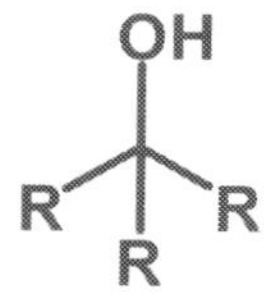


= tertiary alcohol (3° alcohol)

$X-CH_2R$

= primary alkyl halide (1°)

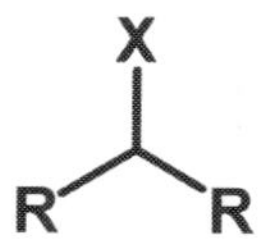


= secondary alkyl halide (2°)

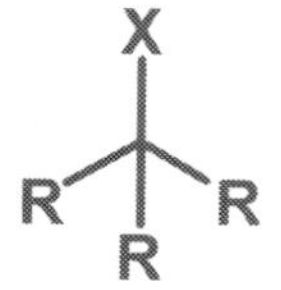


= tertiary alkyl halide (3°)

**Notes:**

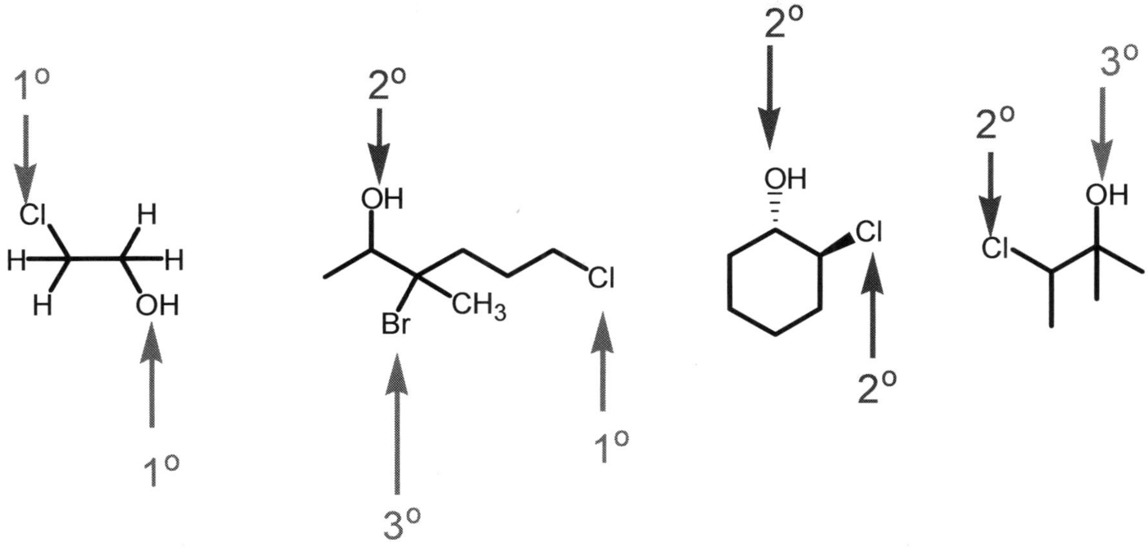
1°
Cl
H
H
H
H
OH
1°
2°
OH
Br
CH3
Cl
3°
1°
2°
OH
Cl
2°
3°
2°
OH
Cl

**Taxol:**

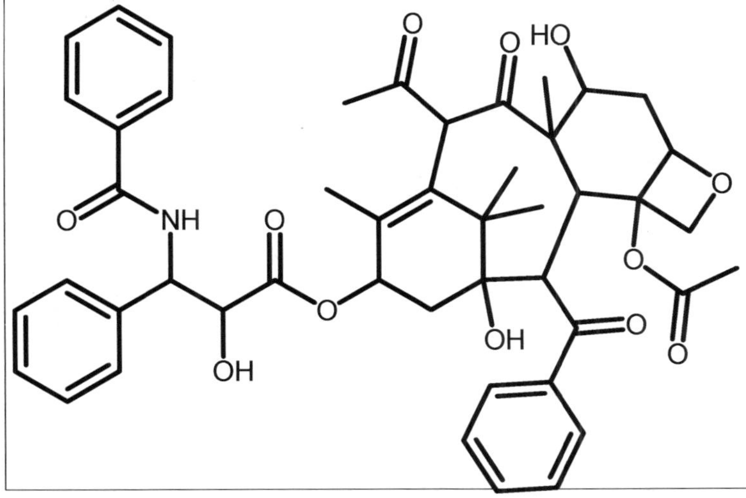
O
NH
O
OH
O
O
O
HO
O
OH
O
O
O

# Lecture Set 3: Conformational Analysis and Reaction of Alkanes

**Suggested Reading:**

**Suggested Problems:**

# 3D Representation of Ethane

Ethane: two $sp^3$ hybridized carbons bound to one another with 6 hydrogen substituents.

(A)

We know that $sp^3$ atoms have a tetrahedral geometry. Drawing one of the carbons in the correct 3D structure:

(B)

Drawing both of the carbons in the correct 3D structure:

**Build ethane with your model kit.**

(C)

**Notes:**

# 3D Representation of Ethane

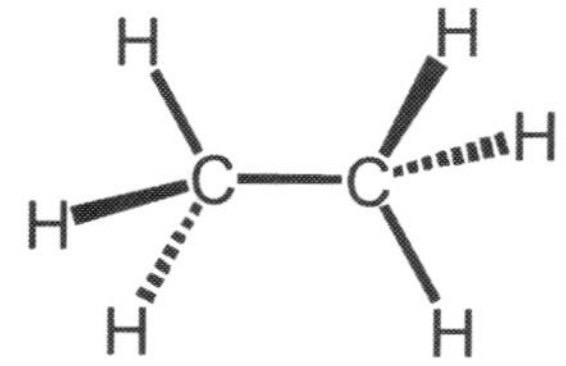


We have two tetreahedra sharing one bond. This has important ramifications for the 3D structure. We can think of this geometrically by joining two tetrahedra at a corner:

the eye sees:

The two triangles are

one another.

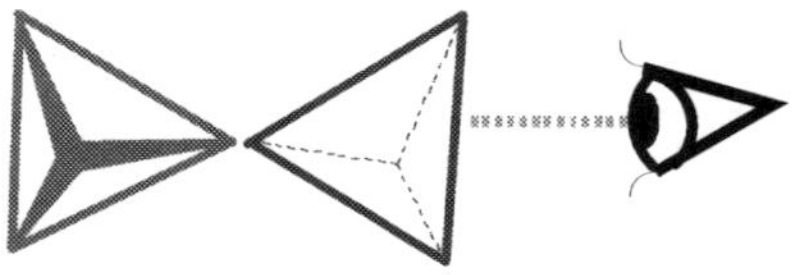

the eye sees:

The two triangles are

**There are different ways the two tetrahedra can be oriented with respect to one another.**

**Notes:**

# 3D Representation of Ethane

Of course, we learned that the dash-wedge representation is used for 3D representation of molecules, so we can also use that to see that there is more than one conformation possible:

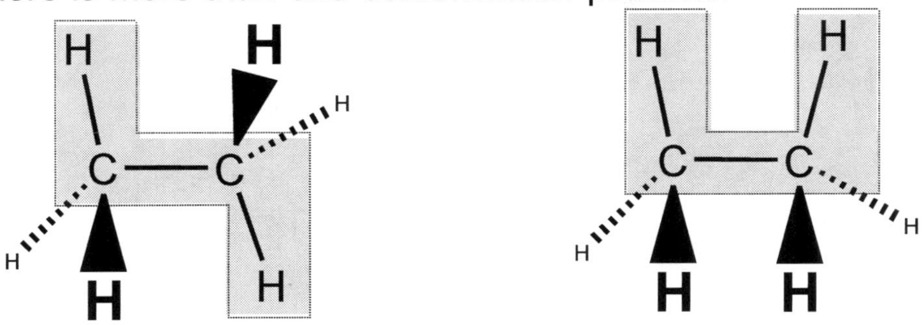


Remember that the 'regular' lines represent bonds in the plane of the paper. I emphasize this here with the yellow shading (and dashed lines for black and white printouts). Notice in the left conformer the two H's coming out at us are far apart from one another, while in the right conformer they are right next to each other. These are two conformational isomers -

**Notes:**

# 3D Representation of Ethane

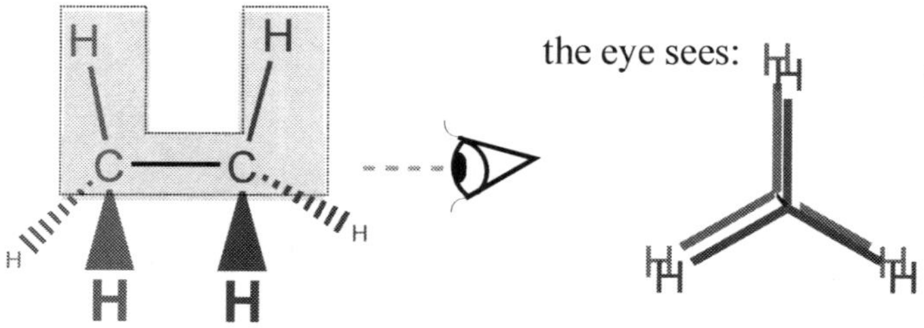


(A) This conformer of ethane is called the

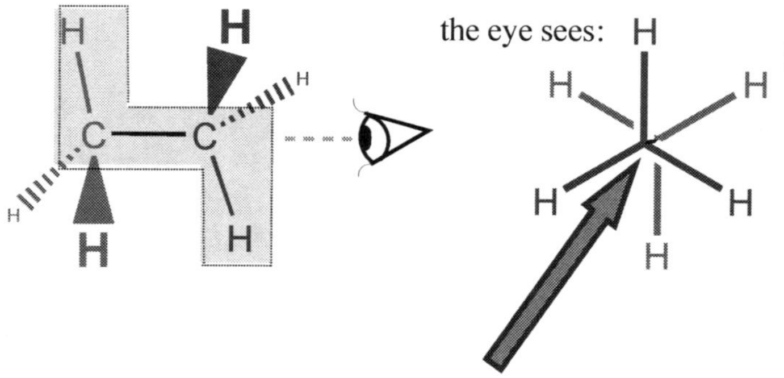


(B) This conformer of ethane is called the

These representations emphasize the fact that we have two different conformers, but are a bit messy, especially at the middle. For this reason, another way of representing this has been developed …

**Notes:**

Put a disk in to block the view of the middle portion of the back, so that the middle is more tidy, and you get what are called

(shown on the right).

Ethane

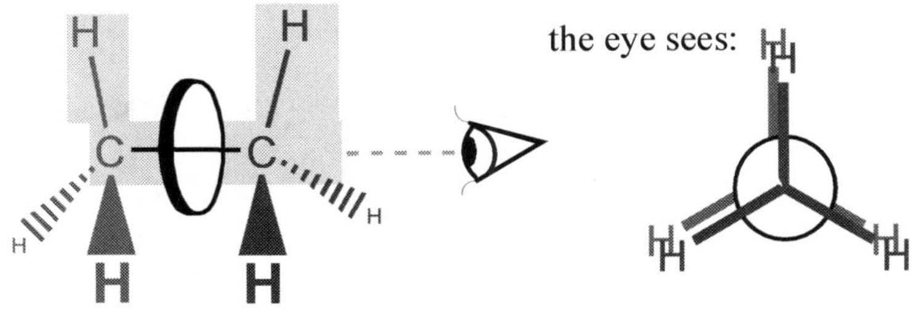


Ethane

Ethane

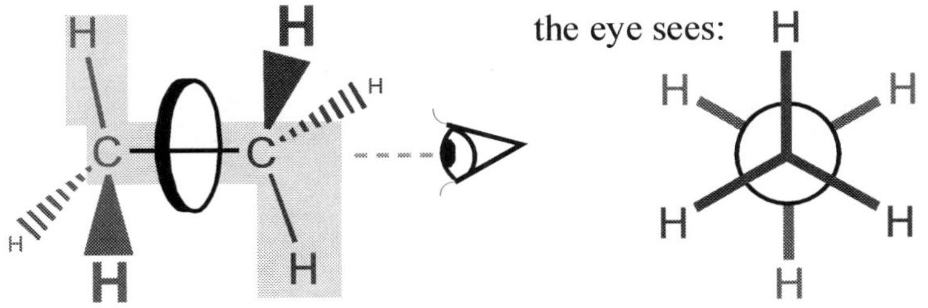


Ethane

If you cut a little circle of paper, put in on your model kit ethane, and look from the end, you'll see exactly what the Newman Projection represents.

**Notes:**

# Stability of Ethane Stereoisomers: Sterics

**Eclipsed** **Staggered**

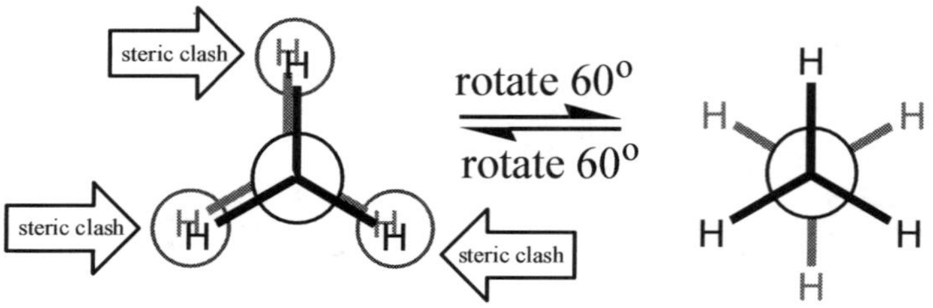


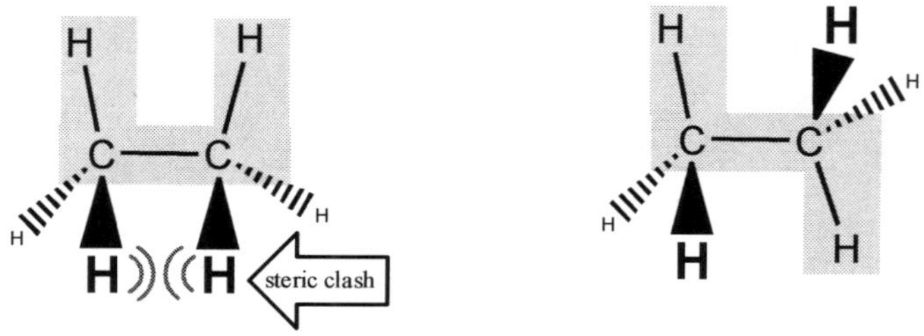


In the eclipsed conformation, H substituents are physically bumping into one another; such interactions are called **steric** interactions (steric interactions lead to different stereoisomers). In staggered, the H atoms are far apart. Because of sterics,

**Notes:**

# Stability of Ethane Stereoisomers: Sterics

Steric clashes between hydrogens in the eclipsed form leads to an energy barrier ($E_a$) to rotation about the C-C bond. This energy barrier is called the **torsional strain** (resistance to twisting). Below is a reaction energy profile showing the energy of the eclipsed and staggered conformers with respect to one another. Note that (A) ____________

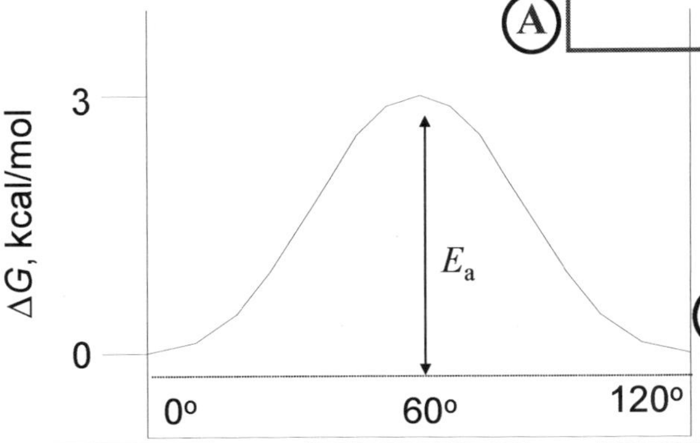


When we **exam**ine this, we notice that the eclipsed conformation corresponds to an energy maximum in the reaction profile, and (B) ____________ between the two staggered conformations, which lie at energy minima.

(C) → (D) → (E)

**Notes:**

# 3D Representation of Butane

Ethane is simple because we only have hydrogens on the C-C bond. Now we look at a more complicated example, butane.

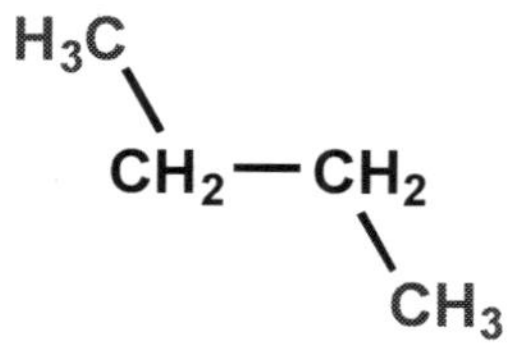

Butane has both hydrogen and methyl groups (blue) attached to the C2-C3 bond. This leads to more possible stereoisomers than we saw with ethane …

**Ethane** **Butane**

**Notes:**

# 3D Representation of Butane

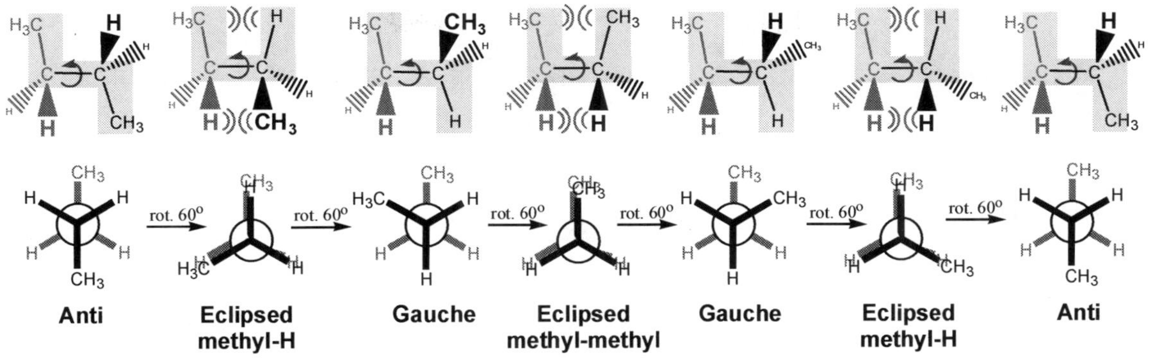


In this scheme, we are holding the back portion of the Newman Projection constant and just rotating the front group in 60° increments. Note that instead of just two conformations like we saw with ethane, we have four distinct conformers:

We need to evaluate steric interactions to determine the relative stabilities of the conformers. Because it is easier to see all of the steric interactions using Newman Projections, let's take a look at these for the four stereoisomers.

**Notes:**

# Stereoisomers of Butane

Ⓐ

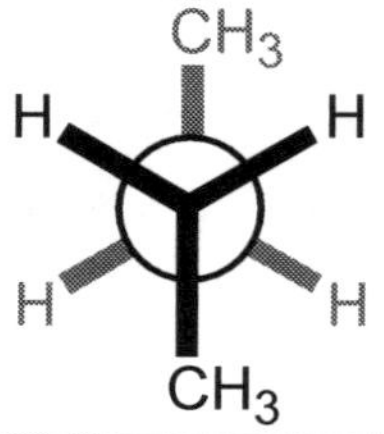


Ⓑ

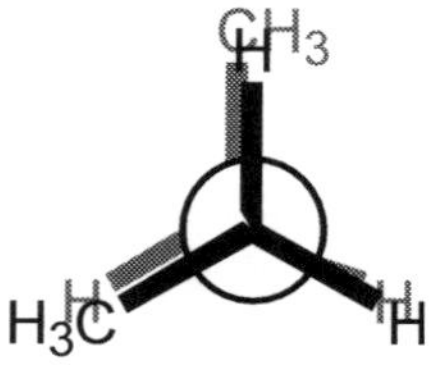


Ⓒ

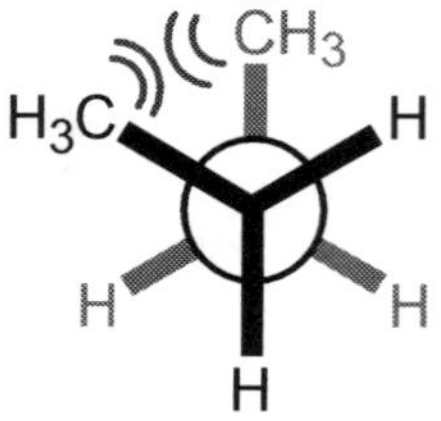


Ⓓ

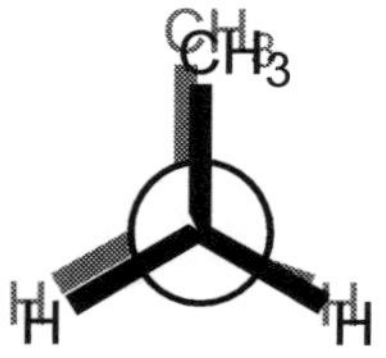


There are no notable steric clashes in this conformation.

The methyl groups are pointing 180° away from each other.

Eclipsing interactions:

Ⓔ

No eclipsing interactions, but the two methyl groups are near one another ; the methyl groups are large enough to bump into one another slightly in this conformation.

Ⓕ

Eclipsing interactions:

Ⓖ

Gauche: (Fr., Teut.) *adj.* Awkward, clumsy

**Notes:**

# Stereoisomers of Butane

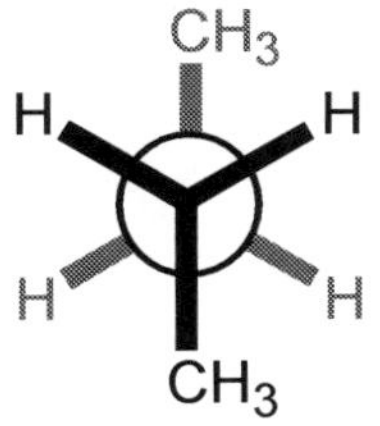


**Anti**

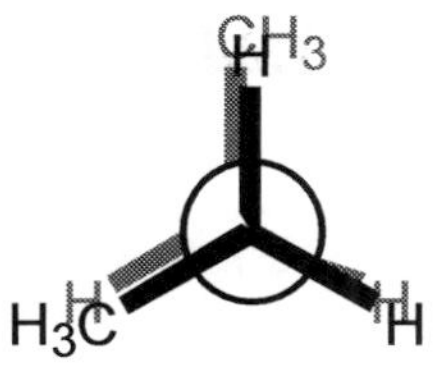


**Eclipsed methyl-H**

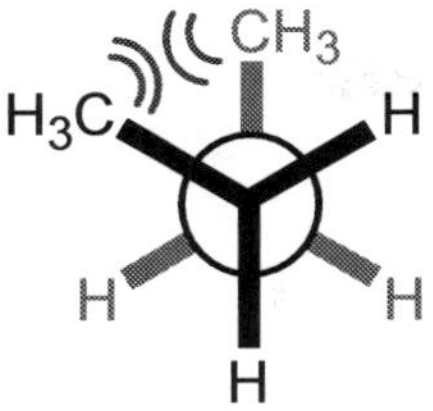


**Gauche**

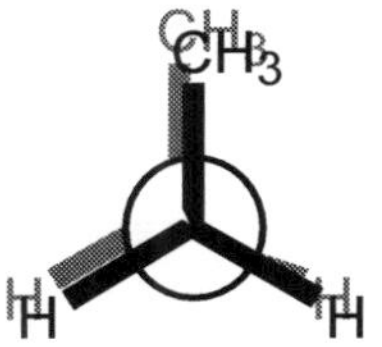


**Eclipsed methyl-methyl**

Since methyl groups are larger than hydrogens,

Ⓐ

In fact, two methyl groups do not have to be eclipsed to interact with one another; there will be an interaction when they are 60° apart from one another, called a

Ⓑ

This interaction is not as strong as when the methyl groups are eclipsed, but this steric interaction does contribute to the energy of the gauche conformation

Ⓒ

**Notes:**

# Conformer Stability

Let's **exam**ine the energetics. Notice that the gauche conformation corresponds to

(A) ______________________

and is therefore an

(B) ______________________

between the two eclipsed isomers.

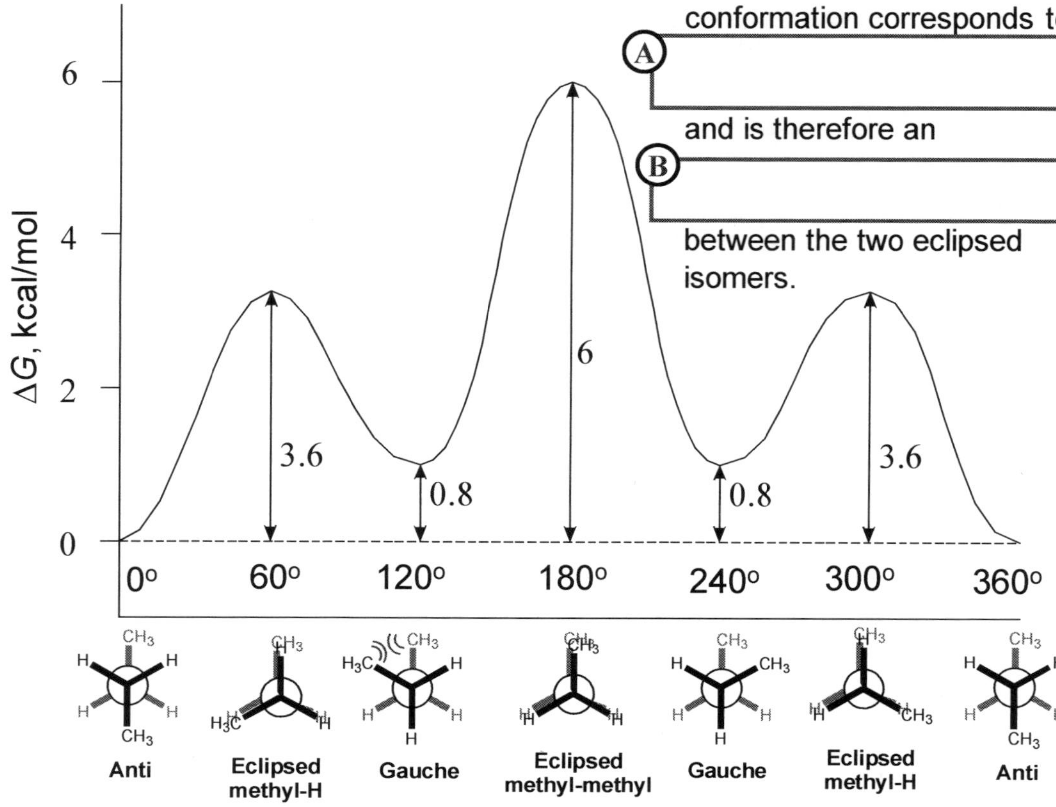


**Notes:**

# Cycloalkane Stability: Ring Strain

We have discussed steric strain (including torsional strain) in linear alkanes. Now we will address the possible strains present in cycloalkanes to help us gauge their stability. If eclipsing or other steric interactions are present, this will drive up the energy, as we saw for linear alkanes.

One additional consideration for cyclic compounds is **angle strain**. Cycloalkanes are made of $sp^3$ hybridized carbons, which favor forming tetrahedral geometries, with bond angles of 109.5° for a perfect tetrahedron. However, check out the angles in some regular polygons:

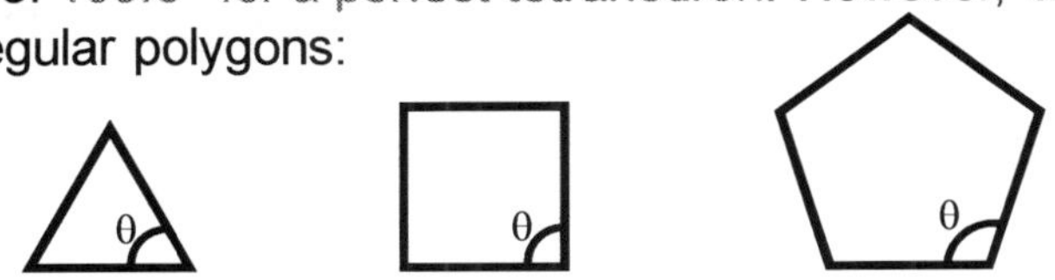


$\theta = 60^{\circ}$ $\theta = 90^{\circ}$ $\theta = 108^{\circ}$

Because the angles in cyclopropane and cyclobutane deviate from the ideal value, there is a good deal of angle strain in these molecules. Cyclopentane and higher cycloalkanes do not have as much angle strain.

**Ring strain:**

**Notes:**

# Angle Strain

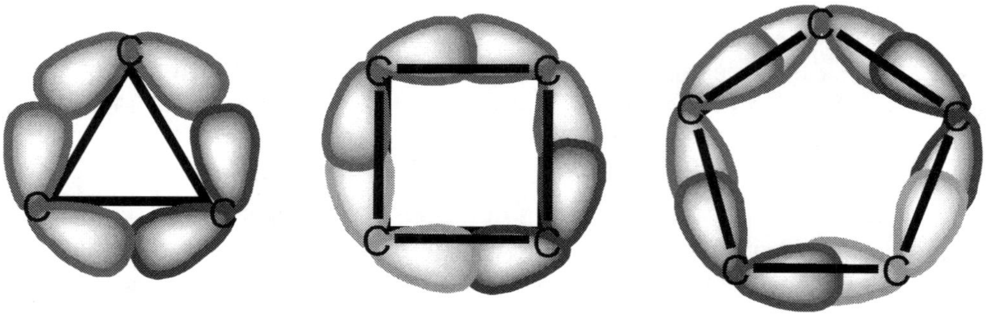


The **origin of angle strain**

Now we consider the other contributors to ring strain: steric interactions (and torsional strain)

**Notes:**

# Ring Strain in Cyclopropane

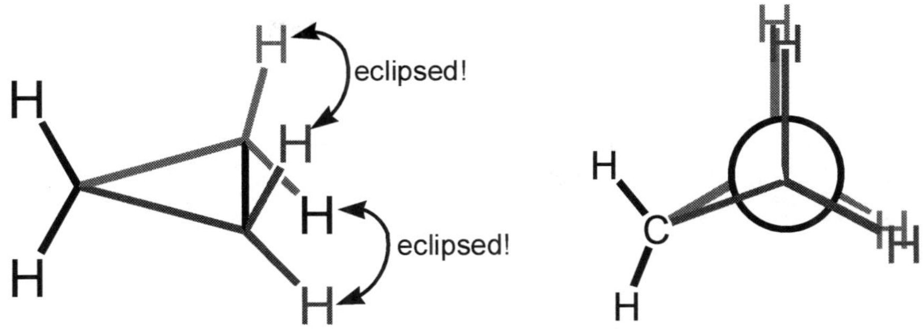


In addition to lots of **angle strain**, cyclopropane also has

The total **ring strain** is about 30 kcal/mol (10 kcal per methylene).

**Notes:**

# Ring Strain in Cyclobutane

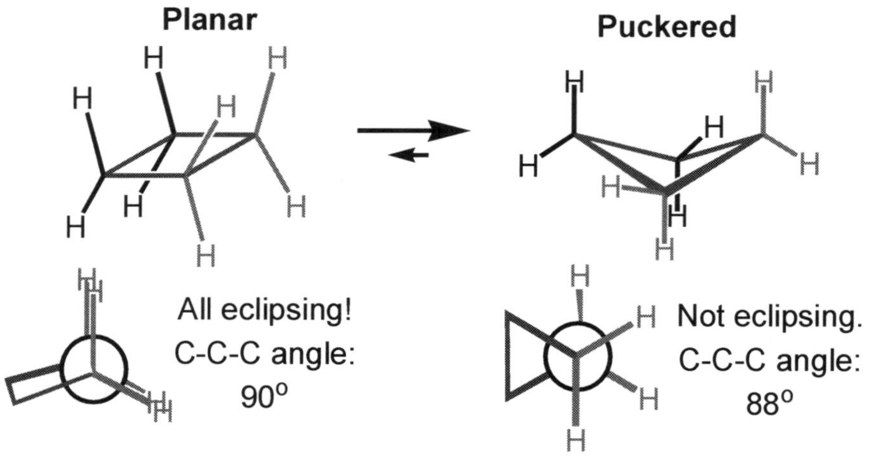


In addition to **angle strain**, planar cyclobutane also has all H's eclipsed, so there is plenty of **torsional strain** as well. **Molecules will always**

In this case, cyclobutane twists ("puckers"), increasing the angle strain a bit, but relieving all of the eclipsing torsional strain. The puckered form, on the right, is the favored form. The total **ring strain** is about 28 kcal/mol (7 kcal per methylene).

**Notes:**

# Cyclopentane

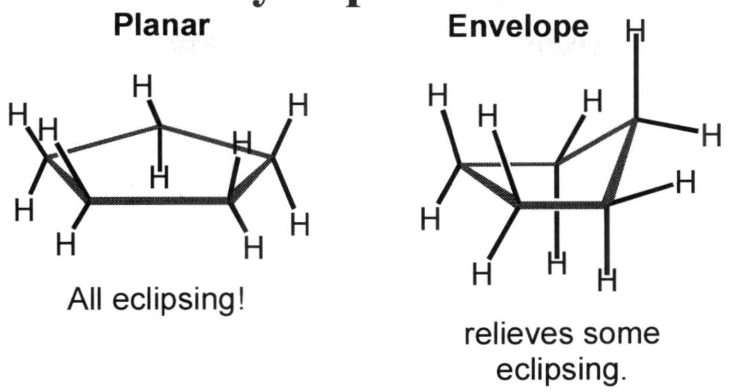


In addition to some **angle strain**, planar cyclopentane also has

(A)

so there is plenty of **torsional strain** as well.
Like cyclobutane, cyclopentane undergoes a distortion to the envelope form, relieving some of the eclipsing torsional strain.

(B)

The total **ring strain** is about 10 kcal/mol (2 kcal per methylene).

**Notes:**

# Cyclohexane

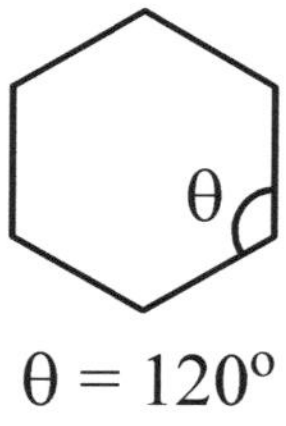


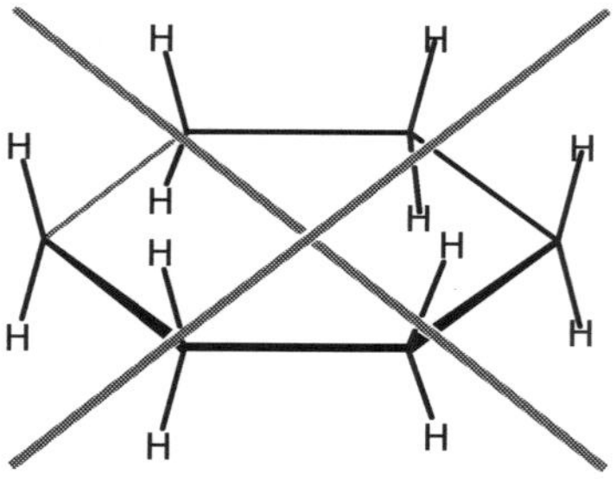

Cyclohexane would have 120° bond angles with all eclipsing hydrogens if it were planar. This is NOT the case. Instead, cyclohexane undergoes distortions, reminiscent of the 'envelope' distortion seen in cyclopentane:

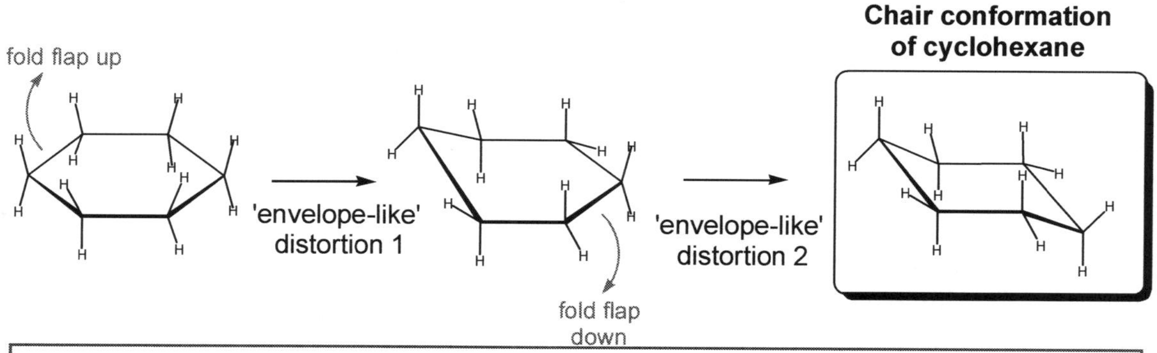


The

We will now study why this is the case.

**Notes:**

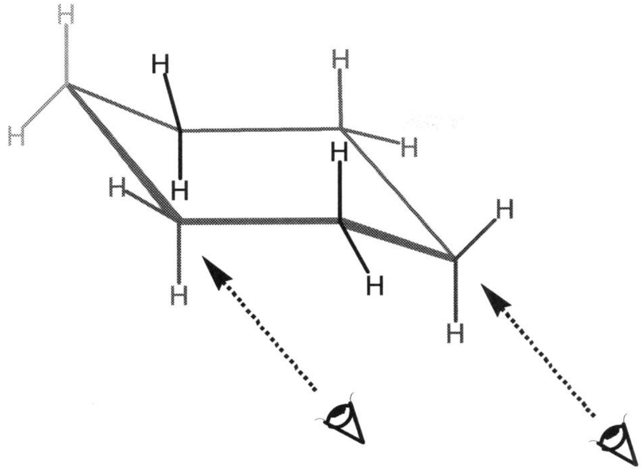


All staggered.

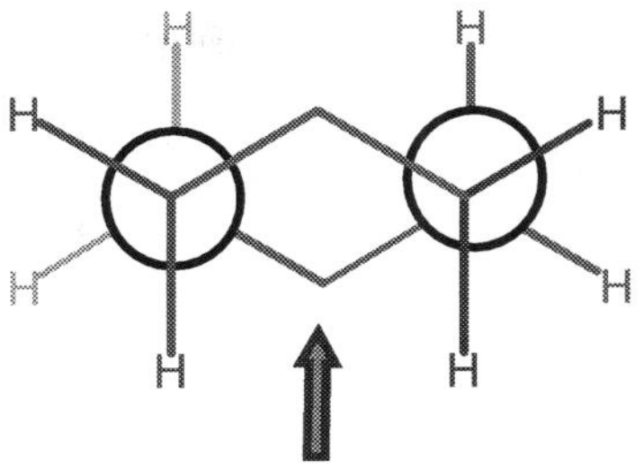


This picture shows two Newman Projections; one projection down the right side and one down the left side of cyclohexane (indicated by the eyes). Build a model to see this more easily.

The chair conformation features
1.

2.

This is **why the chair conformation is very stable**.

**Notes:**

# Ring Strain and Reactions

A reaction that forms a more stable ring is generally more favorable than a comparable reaction that forms a more strained ring.

Which reaction is likely to proceed at a faster rate?

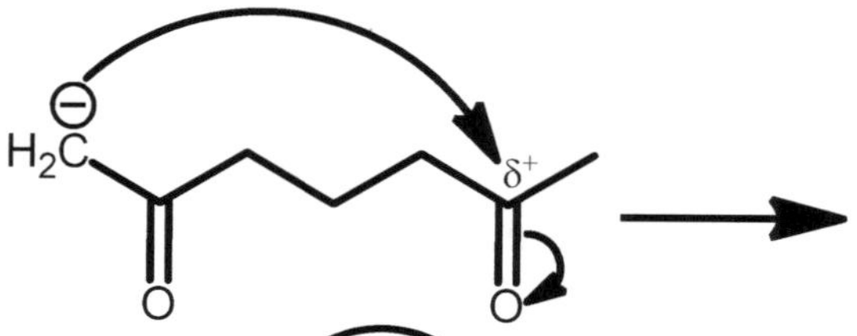


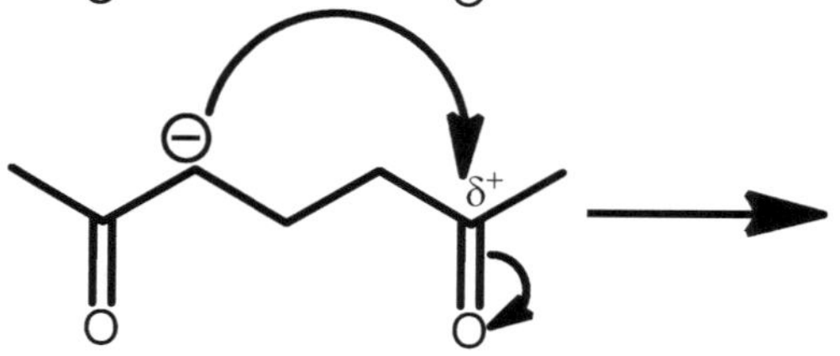


Provide a mechanism by which each species could rearrange to alleviate some ring strain:

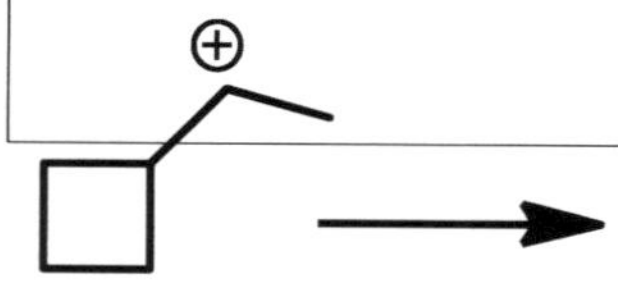

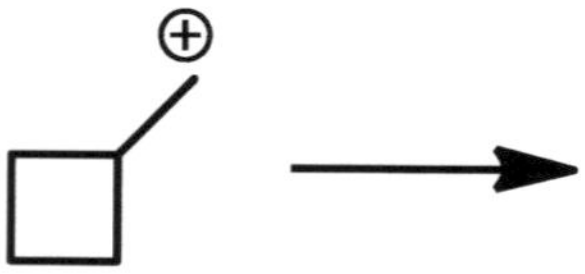

**Notes:**

# Two Types of H Atoms in Cyclohexane

There are two distinct types of hydrogen atoms in cyclohexane. It is very important to understand the difference between these two types.

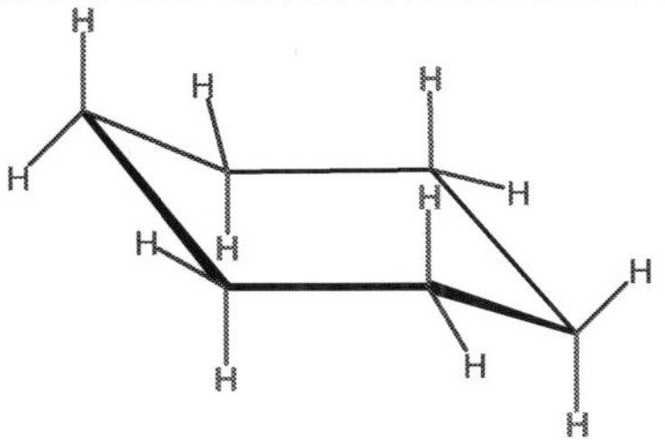

Note: Each C has one equatorial and one axial H and each type alternates up and down around the ring.

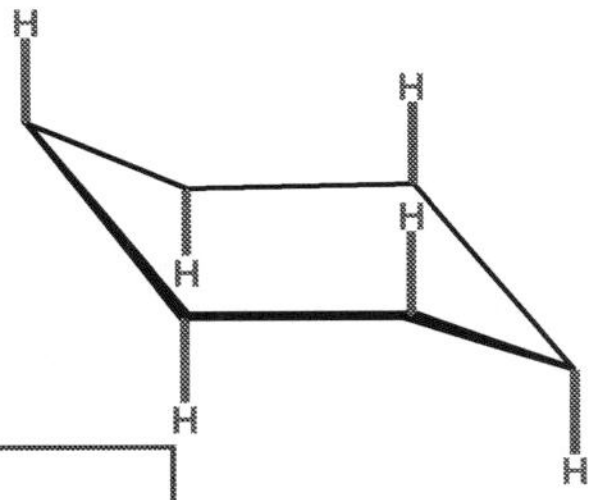

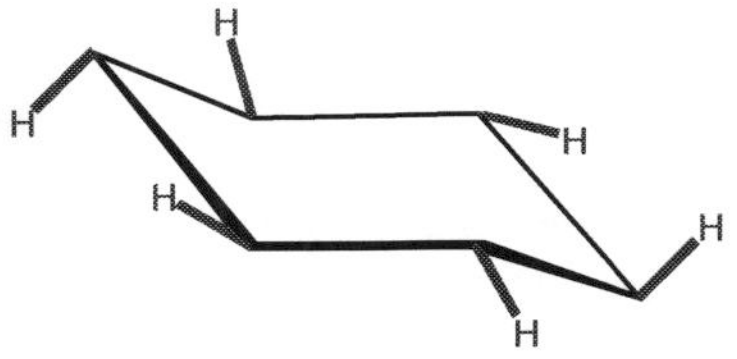

__________: 'stick straight up and down'. If you build a model of the chair conformation of cyclohexane these are the atoms that it will 'stand on.'

__________: point up and down at an angle around the <u>equator</u> of the cyclohexane.

**Notes:**

# A Ring-Flip Switches Axial and Equatorial

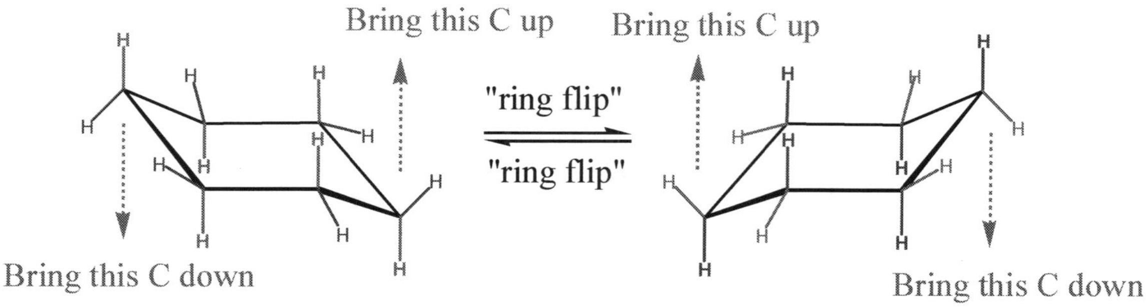


A **ring flip** involves distorting the cyclohexane ring so that the 'up' flap moves to the 'down' position and vice versa. As a result of this process, the equatorial and axial positions are switched around! Try this with a model; it will be much easier to see than envisioning it on paper!

A **ring flip**

**Notes:**

# Differences between Axial and Equatorial: Cross-Ring Interactions

(A)

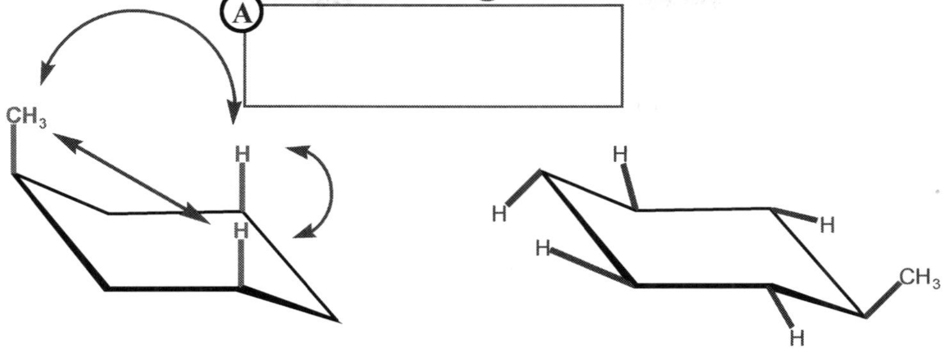


Although cyclohexane has no ring strain, if we begin adding substituents we may introduce some steric strains. If we place a methyl group on the 1-position of cyclohexane in an axial position, it will be close enough to the other axial H's to bump into them (in the 3 and 5 position). This type of steric interaction is called a

(B)

Equatorial positions do not have these interactions, so

(C)

**Notes:**

# Differences between Axial and Equatorial: Configurational Isomers

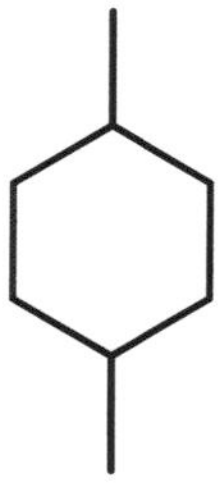

1,4-dimethylcyclohexane

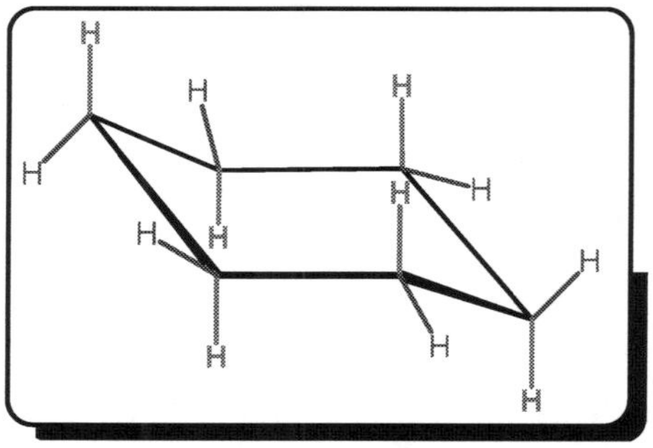

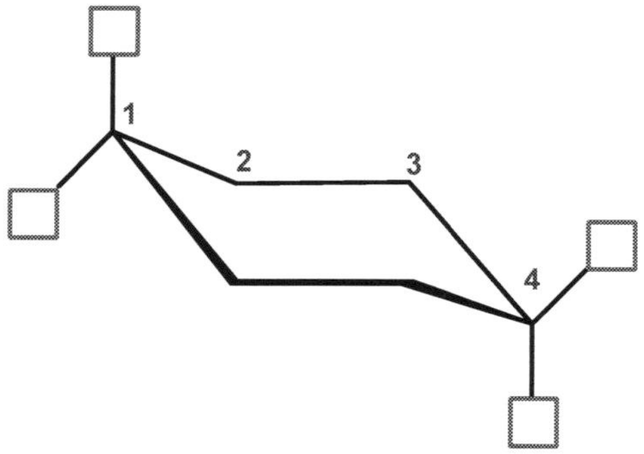


Given a simple substituted cycloalkane, we need to consider how the presence of axial and equatorial sites will affect where we place the substituents. One methyl must go on each of carbons 1 and 4. We have a few possible choices ...

**Notes:**

# Differences between Axial and Equatorial: Configurational Isomers

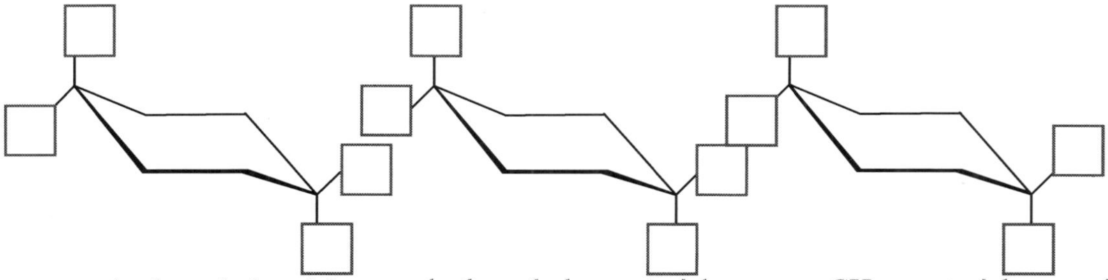

| both methyls axial (one up, one down) | both methyls equatorial (one up, one down) | one $-CH_3$ equatorial, one axial (both up) |
|---|---|---|
| *trans*-isomer | *trans*-isomer | *cis*-isomer |

Depending on where we put the methyl substituents, we have two different **configurational isomers**. A configurational isomer in which substituents are on the same side (both 'up' in this case) is referred to as a ***cis-*** isomer, while a configurational isomer with substituents on opposite sides is referred to as the ***trans-*** isomer.

cis: (L.) *adj.* On this side. i.e.- Cisalpine-on the same side of the alps as Rome.

**Notes:**

Remember that *trans* indicates two substituents on opposite sides of a ring and *cus* indicates two substituents on the same side of the ring; the names DO NOT refer to whether a substituent is axial or equatorial.

# Configurational Isomers

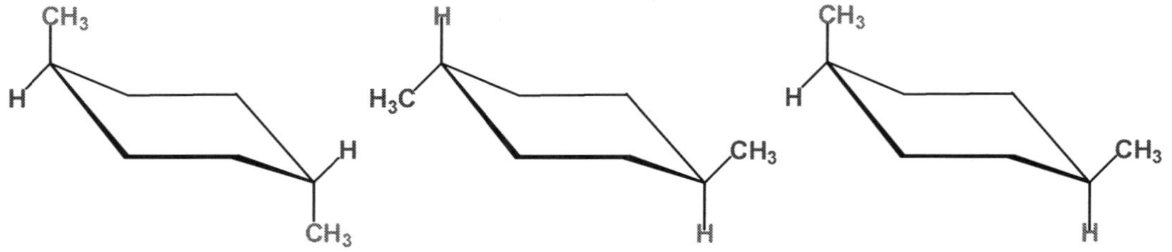


| both methyls axial (one up, one down) | both methyls equatorial (one up, one down) | one $-CH_3$ equatorial, one axial (both up) |
|---|---|---|
| *trans*-isomer | *trans*-isomer | *cis*-isomer |

Because axial methyl groups will have 1,3-diaxial interactions, the *trans*-isomer on the right is the most stable. The *trans*- isomer on the left will ring flip to form predominantly the isomer on the right. In the *cis*- isomer, one methyl is axial and one is equatorial. Ring flipping will result in the same thing, so there is only one possible *cis*- form. When naming the molecules we need to denote which isomer it is. So, we add a *cis*- or *trans*- prefix to the name: (left, middle) *trans*-1,4-dimethylcyclohexane
(right) *cis*-1,4-dimethylcyclohexane

**Notes:**

# Configurational Isomers of Cyclohexane

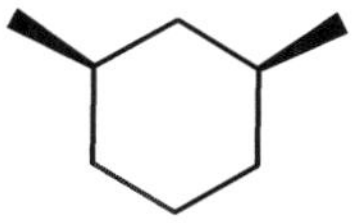

*cis*-1,3-dimethycyclohexane

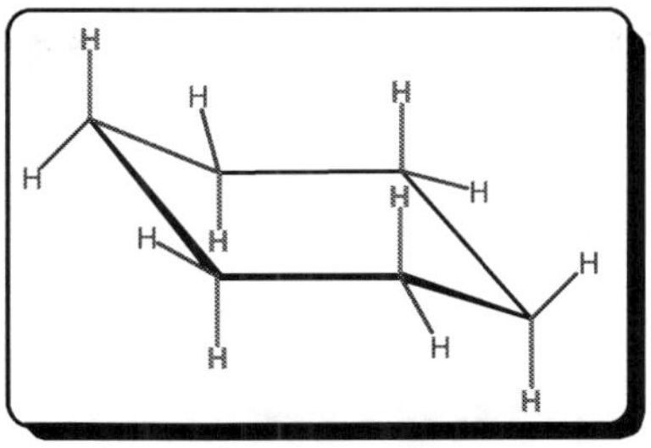


For Reference

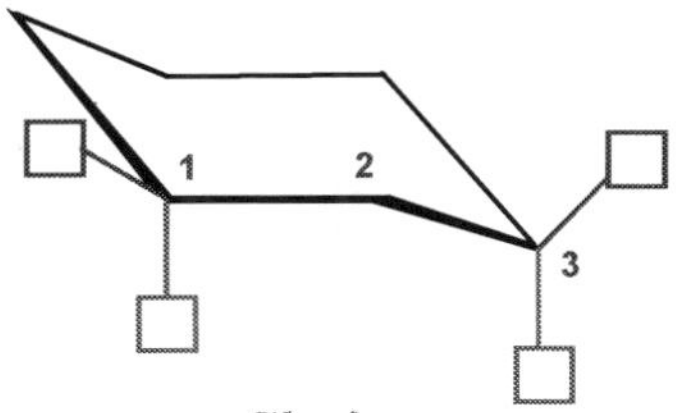


Choices

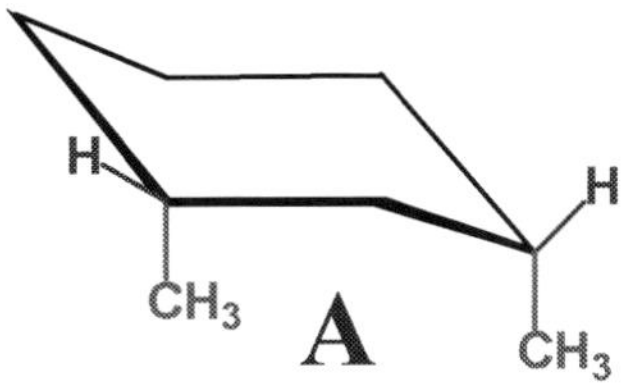


**A**

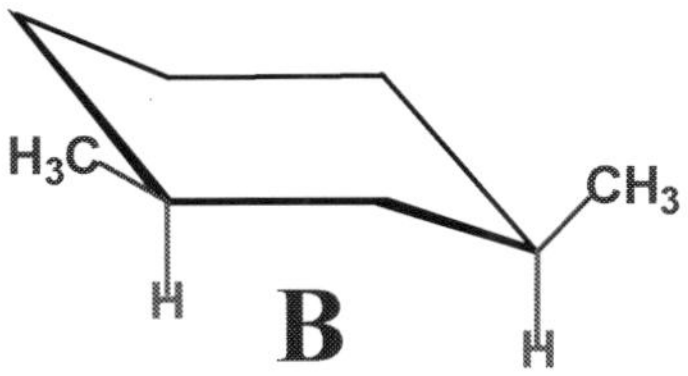


**B**

The structure, as well as the *cis*- prefix in the name, tell us that both methyls must be on the same side. We can choose both axial or both equatorial.

**Notes:**

# Configurational Isomers of Cyclohexane

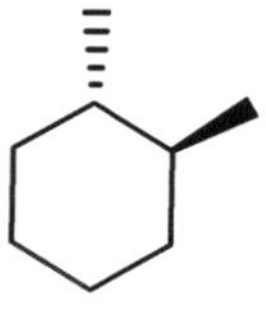

*trans*-1,2-dimethycyclohexane

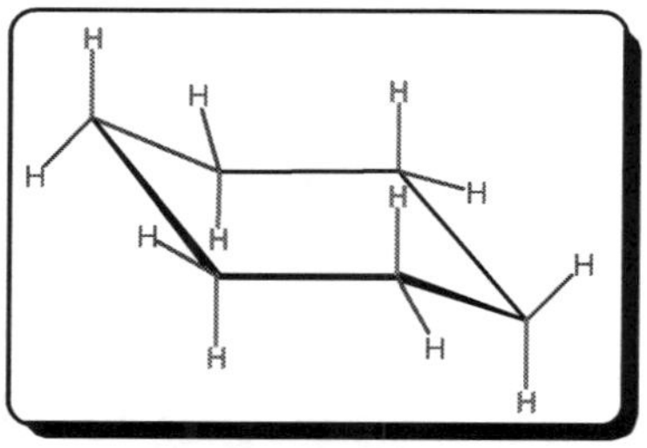


For Reference

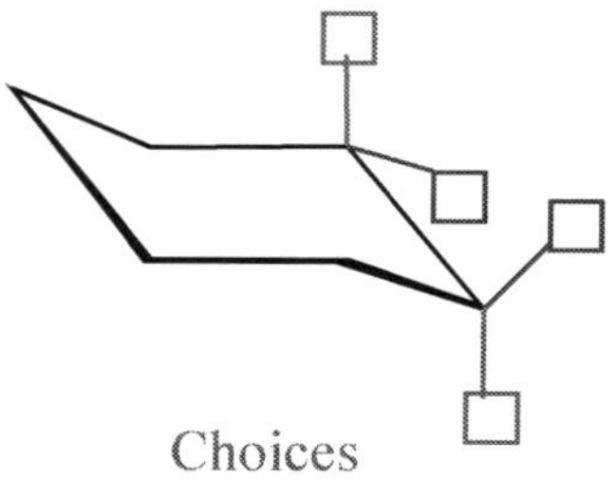

Choices

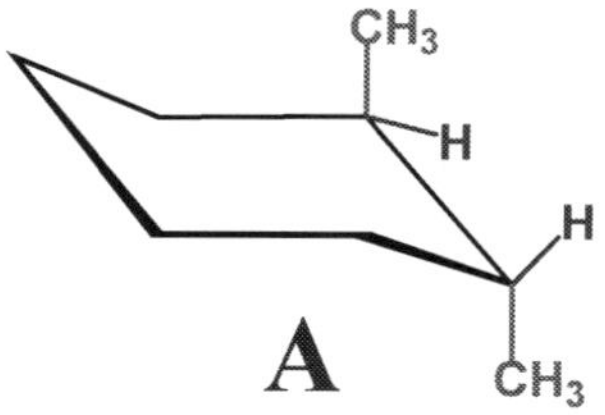


A

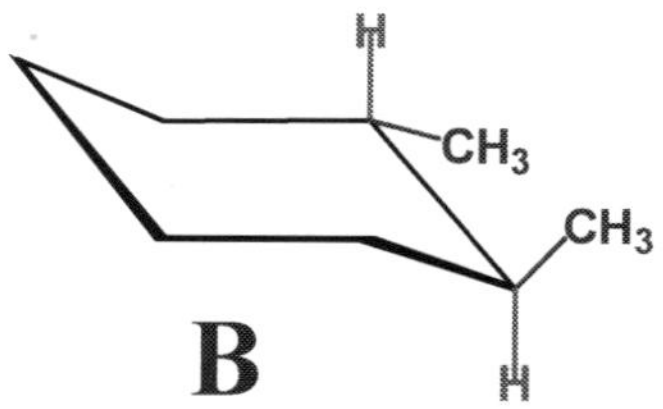


B

Axial methyls will have strong 1,3-diaxial repulsion, so this form will be unfavorable. The most stable form is therefore:
Note that the two equatorial methyl groups in this conformation

**Notes:**

# Configurational Isomer Problem

**Example.** Provide the IUPAC name for each of these structures. How many axial and how many equatorial H atoms are in each?

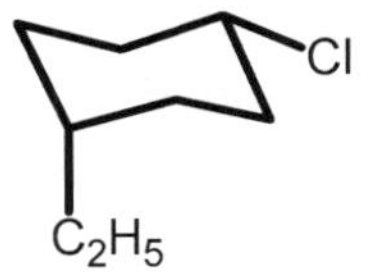


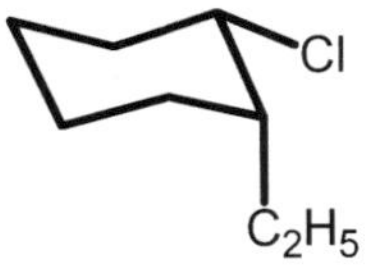


**Example.** Provide the lowest energy structure for each of the following molecules. How many axial and how many equatorial H atoms are in each?

*cis*-1-methyl-2-propylcyclohexane

*trans*-1-methyl-2-propylcyclohexane

**Notes:**

# Other Conformations of Cyclohexane

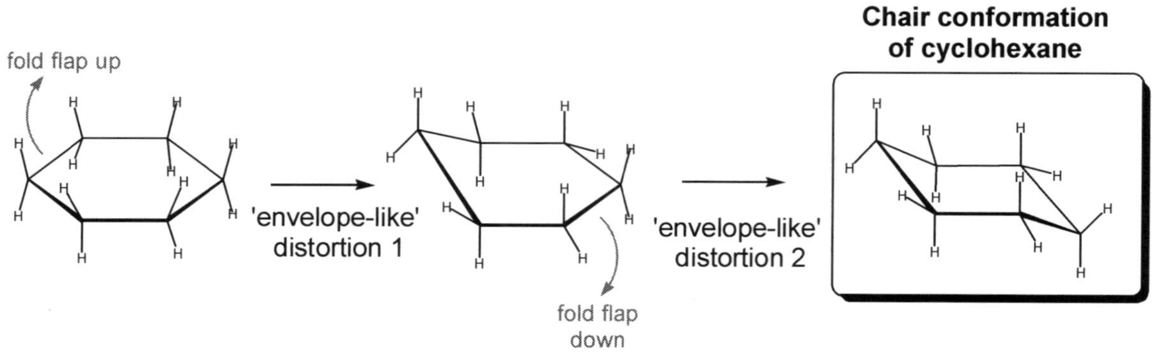


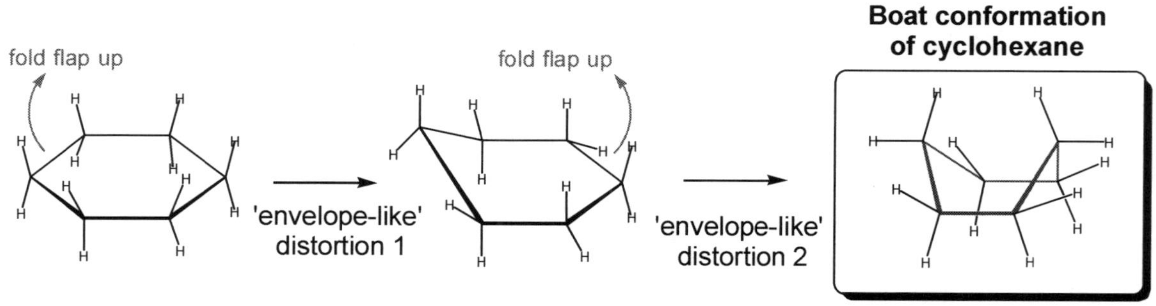


**Notes:**

# Other Conformations of Cyclohexane

Steric interaction
(known as a
**"flagpole" interaction**)

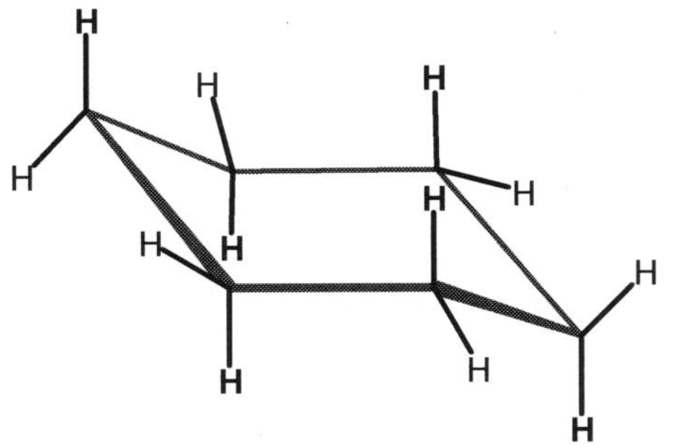

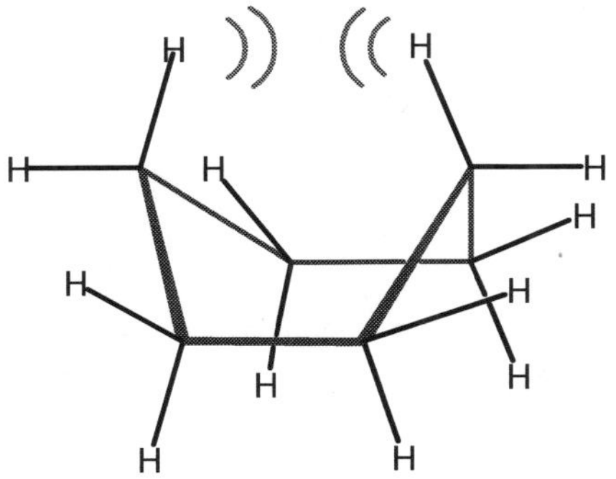

"**chair**" conformation is very stable:

"**boat**" conformation

**Notes:**

# Preparing Alkyl Halides from Alkanes

Alkyl halides can also be prepared from alkanes. Alkanes do not have very polar bonds, and do not readily react with common acids or bases. If we generate a very reactive species, like a radical

we can observe reactions of alkanes. A good way to generate a halide radical is by excitation of $X_2$ with light.

$$R\text{–}H + X_2 \xrightarrow[\text{(light)}]{h\nu} R\text{–}X + HX$$

The dihalides have different reactivities towards radical formation, so only $Cl_2$ and $Br_2$ are commonly used in the above reaction:

$$I_2 < \boxed{Br_2 < Cl_2} < F_2$$

REACTIVITY →

Because we are using such reactive species (radicals) it is often difficult to control these reaction to yield a single product, as we will see …

**Notes:**

# Radical Chain Mechanism

We can write the discrete steps of the radical chain mechanism:

Ⓐ Initiation

Ⓑ Propagation

Ⓒ Net Reaction

Ⓓ Termination

**Notes:**

# Radical Chain Mechanism

... or we can write a simple cycle (shows **initiation** and **propagation** but not **termination**):

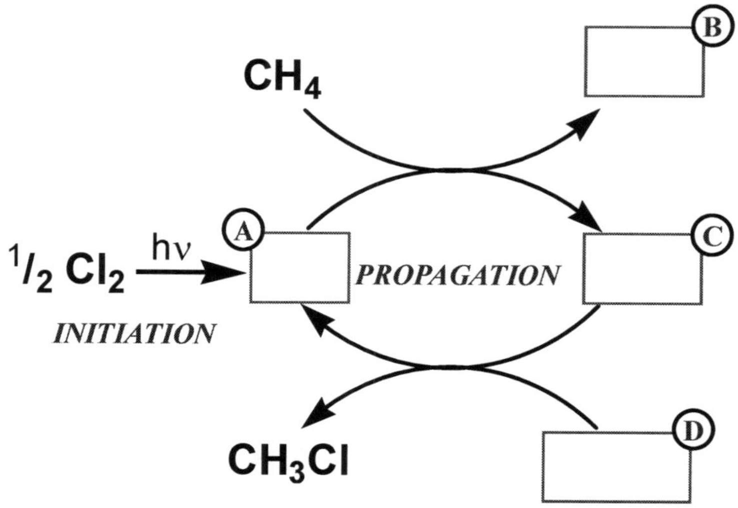


Termination occurs when any two radicals come together.

**Notes:**

# Preparation of RX from RH

We've discussed the divergent reactivity of the dihalides and that only $Cl_2$ and $Br_2$ are commonly used in this radical chain reaction, but the structure of the alkane we use also plays an important role.
The key step is the homolytic cleavage of the alkane C-H bond to form the radical:

R–H → R• + H•   $\Delta H$ = BDE (**bond dissociation enthalpy**)

Easier radical formation ↓

**BDE Trends:**

| Structure | BDE | |
|---|---|---|
| $CH_3$—H | 104 kcal | |
| $CH_3CH_2$—H | 98 kcal | |
| $CH_3CH_2CH_2$—H | 98 kcal | (1°) |
| $(CH_3)_2CH$—H | 95 kcal | (2°) |
| $(CH_3)_3C$—H | 91 kcal | (3°) |

(opposite the trend we discussed for carbanions! Same trend for carbocations).

**Notes:**

# Regioselectivity of Halogenation

Some alkanes give the same product regardless of which H is removed upon radical formation:

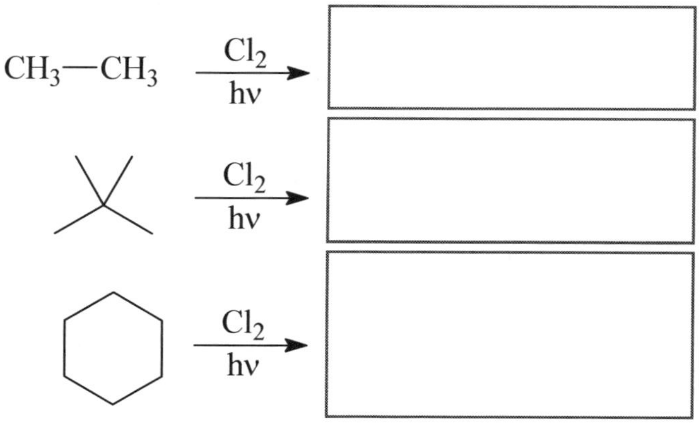


In some cases, we have two or more possible H atoms that can be displaced when we form the alkyl radical. Consider the following example ...

**Notes:**

# Regioselectivity of Halogenation

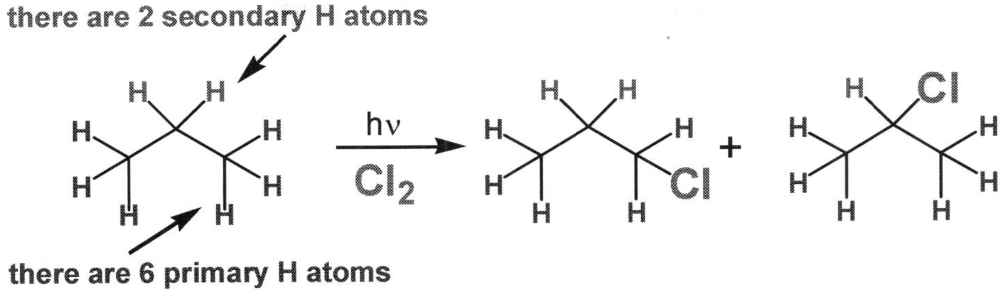


| | | |
|---|---|---|
| Experimentally observed: | 43% | 57% |
| Statistically anticipated: | 75% | 25% |

We observe a greater than statistically anticipated reactivity at the secondary site. The reaction has a greater regioselectivity for this site. We already know that the secondary radical is more stable than the primary, and that the secondary C-H bond has a lower BDE, so this regioselectivity makes sense.

**Notes:**

# Regioselectivity of Halogenation

When $Cl_2$ is used, the relative reactivity can be approximated as:

| **Relative Reactivity of C–H:** | 3° | 2° | 1° |
|---|---|---|---|
| | 5.2 | 3.9 | 1 |
| | (5) | (4) | (1) |

Use this to predict product distribution for the chlorination of propane:

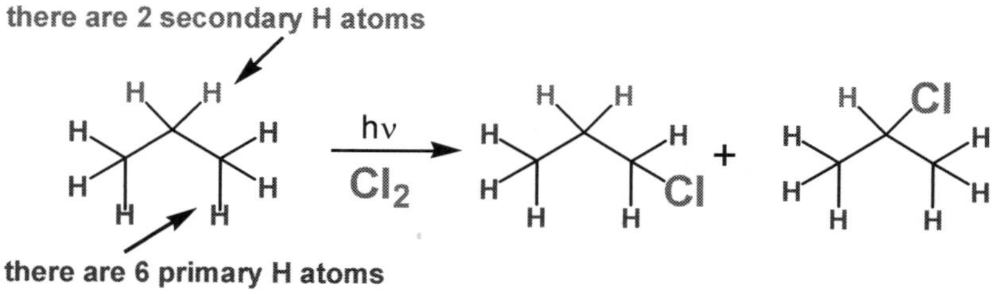


Rel. amount of 1° product = reactivity of 1° H x number of 1° H's

Rel. amount of 2° product = reactivity of 2° H x number of 2° H's

= 3.9 x 2 = 7.8 = 43% primary

= 1 x 6 = 6 = 57% secondary

**Notes:**

# Regioselectivity Problem

Example. Show the products of this reaction and indicate what relative percentage of each product you would expect based on relative reactivity estimates.

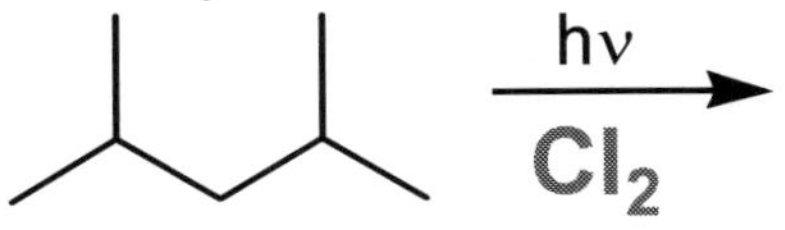


**Notes:**

# Regioselectivity Answer

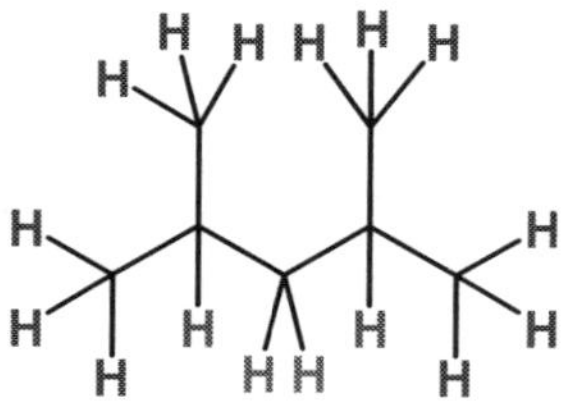


**there are 12 primary H atoms**
**there are 2 secondary H atoms**
**there are 2 tertiary H atoms**

The three possible products:

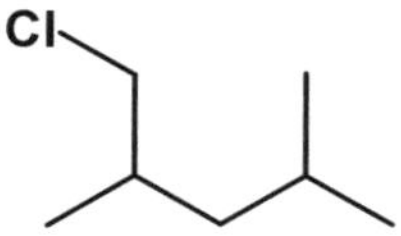


**Primary Alkyl Chloride**

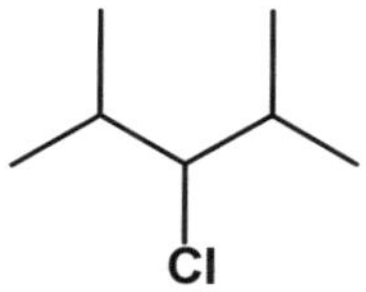


**Secondary Alkyl Chloride**

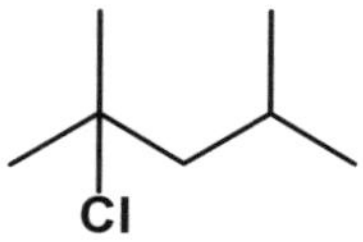


**Tertiary Alkyl Chloride**

To get the product ratios we use the formula (number H's x reactivity):
Primary: 12 x 1 = 12
Secondary: 2 X 3.9 = 7.8
Tertiary: 2 x 5.2 = 10.4
(total = 12 + 7.8 + 10.4 = 30.2)
Percent: 12/30.2 x 100 = 39.7% 7.8/30.2 = 25.8 10.4/30.2 = 34.4%

NOTE: No calculator on the exam (see syllabus).

**Notes:**

# Regioselectivity of Halogenation

Bromine is less reactive and therefore more **selective** than chlorine. Consider the observed products for these reactions:

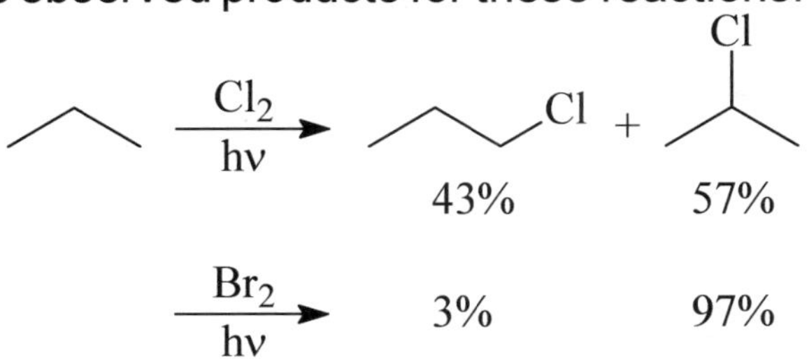


From these data we can see that the selectivity of bromine must be more pronounced than in for chlorine:

| 3° | 2° | 1° |
|---|---|---|
| 1640 | 82 | 1 |
| (2000) | (100) | (1) |

We can estimate product distributions in the same way as we did with chlorine.

**Notes:**

# Regioselectivity of Halogenation

Since bromine exhibits greater regioselectivity, we get a 'cleaner' reaction, one in which we get predominantly one product, and the bromination reaction is therefore more useful synthetically in many cases.

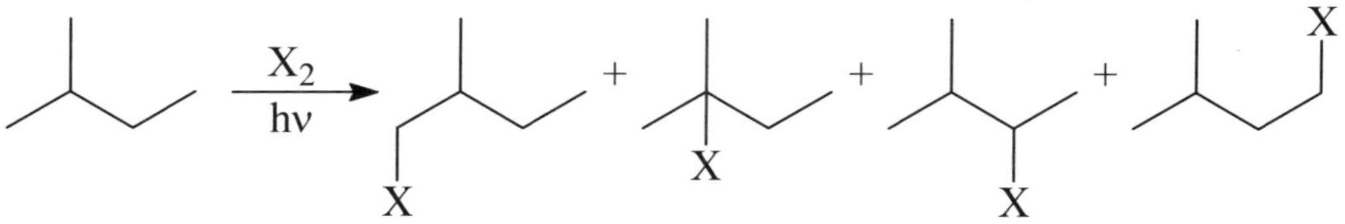


| | | | | |
|---|---|---|---|---|
| $Cl_2$: | 28% | 23% | 35% | 14% |
| $Br_2$: | ~0% | 90% | 9% | ~0% |

**Notes: TAKE HOME LESSON:**

Bromination is very selective and the major product thus tends to be one resulting from replacing the easiest-to-remove H (the one that leaves behind the most stable radical) with a Br.

# Polycyclic Hydrocarbons

## Bicyclic Alkanes

## Tricyclic Alkanes

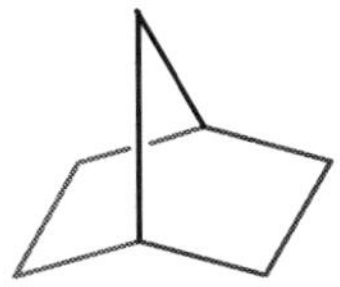

Bicyclo[2.2.1]heptane

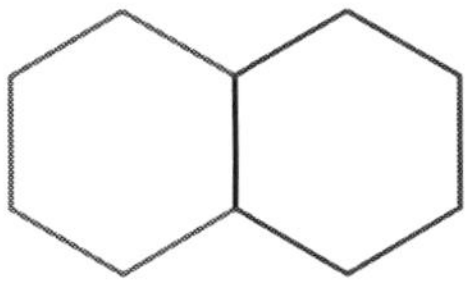

Bicyclo[4.4.0]decane

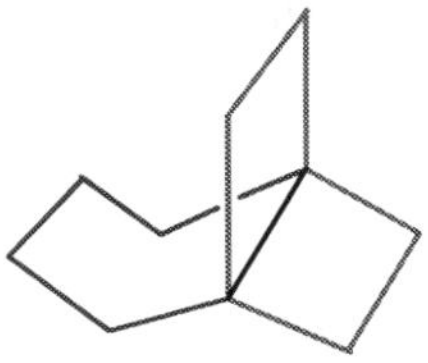

Tricyclo[4.2.2]decane

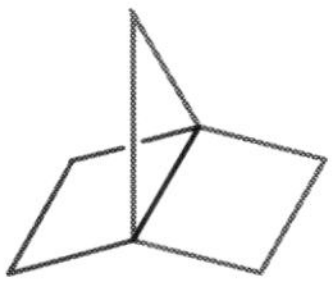

Tricyclo[2.2.1]heptane

**Nomenclature:** polycyclic alkanes are named as "**bicyclo[x.y.z]alkane**" or "**tricyclo[x.y.z]alkane**" (depending on whether 2 or three rings are present), where x, y, and z are the number of carbons joining the shared vertex (called a bridgehead) and $x \geq y \geq z$ (put biggest number first). Start numbering at a bridgehead. Number the largest ring first, ending with the smallest ring.

**Notes:**

# Lecture Set 4:
# Nomenclature, Properties and Synthesis of Alkenes

**Suggested Reading:**

**Suggested Problems:**

# Alkenes versus Alkanes: Unsaturation

Alkanes: $C_nH_{2n+2}$ **saturated** hydrocarbons
Alkenes: $C_nH_{2n}$ **unsaturated** hydrocarbons (put a double bond in an alkane)
(olefins)

Each removal of two hydrogens from a saturated hydrocarbon (alkane) is one degree of unsaturation (can add one $H_2$ molecule to it)

A **degree of unsaturation** can be:

(A)

A general formula to determine the degree of unsaturation of a molecule from its formula ($C_cH_hN_nO_oX_x$, where X = F, Cl, Br or I) is:

(B)

$$\text{Degree of unsaturation} = 1 + c - \frac{\_\_\_\_}{2} - \frac{\_\_\_\_}{2} + \frac{\_\_\_\_}{2}$$

*Note that oxygen's presence does not impact the degree of unsaturation

**Notes:**

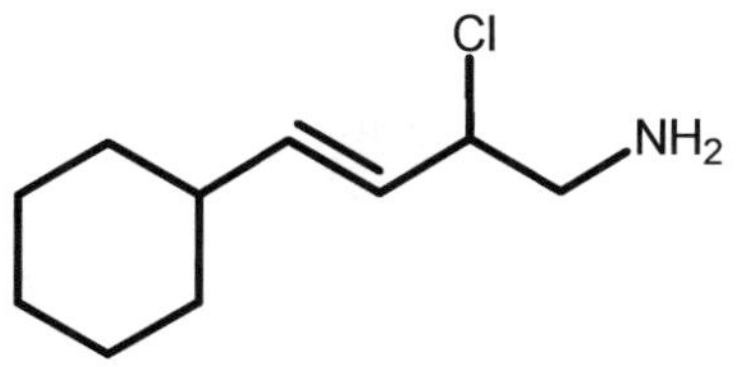


$C_{10}H_{18}ClN$

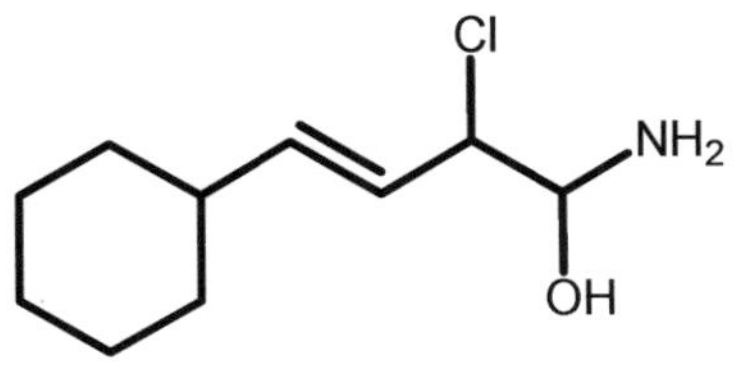


$C_{10}H_{18}ClNO$

# Degrees of Unsaturation

If you learn a compound's molecular formula by doing an analysis of some kind, you can use the information to help determine what the molecule is.

Example: For each of the following molecular formulae, list all functional groups (selecting from those provided in Lecture Set 2) that *could* be present in the molecule.

$C_6H_{12}O$

$C_6H_{14}O$

$C_6H_{12}O_2$

$C_6H_{14}O_2$

**Notes:**

# Naming Alkenes

Nomenclature of alkenes is related to that for alkanes, with modifications:

1. always designate the longest chain having

(A) [ ] as the parent chain.

2. Number the parent chain such that the double bond

(B) [ ]

4. change the 'ane' ending of the name you would use if it were an alkane

(C) [ ]

5. Tell where the double bond starts:

(D) [ ]

**Notes:**

# Naming Alkenes

propene

1-butene

2-butene

1-pentene

2-pentene

1-hexene

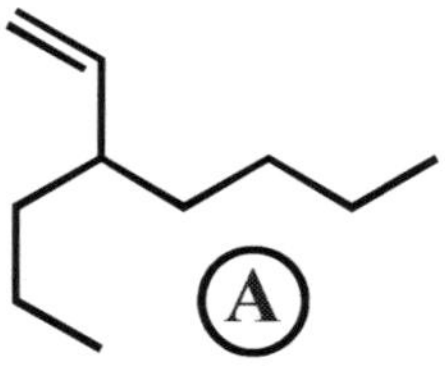

A

**Notes:**

4. If there is more than one double bond, use a "-diene", "-triene", etc. at the end of the name in place of the "ene", with multiple numbers indicating where the double bonds are:

(A)

(B)

1,4,8-nonatriene

5. If an alcohol is present, it has higher priority than the alkene functionality, and the molecule has an "-ol" ending (with a number to tell you where the OH is!) after the "ene" suffix (change "ene" to "en"), and the **alcohol is given the lowest number**:

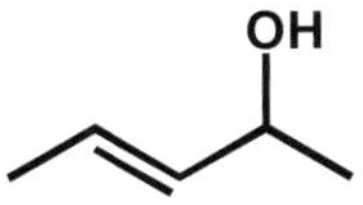


3-penten-2-ol

(C)

4-penten-1-ol

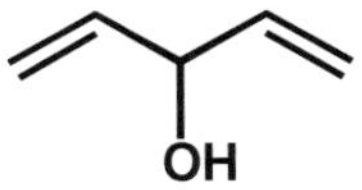


(D)

**Notes:**

# Alkene Isomers: *cis-* or *trans-*

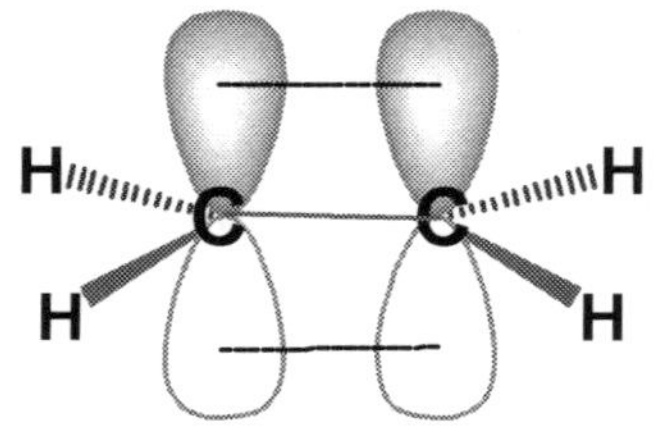


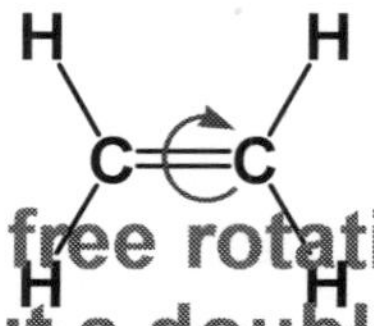


**NO free rotation about a double or triple bond**

(A)

(B)

*cis*-alkene

(C)

*trans*-alkene

As we saw for cycloalkanes, when we have two <u>configurational isomers</u>, the one with two substituents on the same side is the *cis* isomer; the one with substituents on opposite sides is the *trans* isomer. The same holds for alkenes which have **TWO** non-hydrogen substituents.

**Notes:**

# Alkene Isomers: *cis-* or *trans-*

Ⓐ  ______-2-hexene

Ⓑ 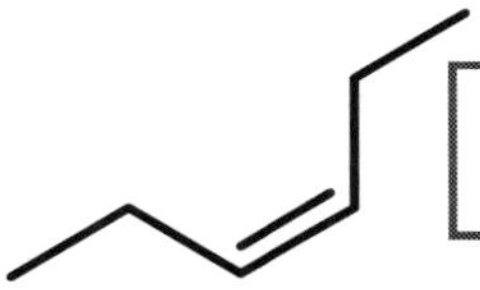 ______-3-hexene

Ⓒ 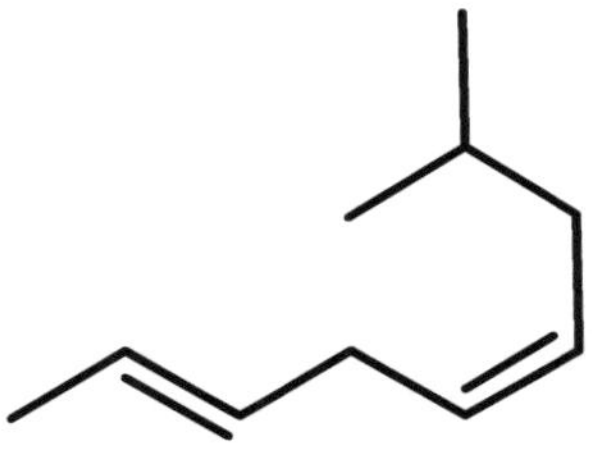 ______-8-methyl-2,5-nonadiene

Ⓓ

**Notes:**

# What about trisubstituted?

(A) cis

(B) trans

(C)

(D) The King

(E) The Return of the King

***Z-***

***E-***

**Notes:**

# Alkene Isomers: *E*- or *Z*-

Is this alkene the *cis*- or the *trans*- isomer?

(A) 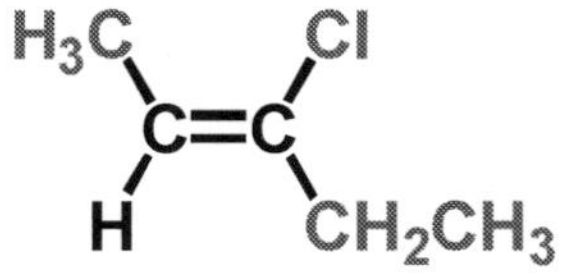


For tri- or tetra-substituted alkenes we use *E*- (from German 'entgegen', opposite) or *Z*- (from German 'zusammen', together) notation:

If the two **highest priority groups** are

| | |
|---|---|
| on same side, then | (B) |
| opposite sides, then | (C) |

| (D) | (E) |
|---|---|
| ***Z*-isomer** | ***E*-isomer** |

we need to know how to determine the priority of each substituent …

**Notes:**

# Priority Rules

**Priority Rules (**Cahn-Ingold-Prelog Rules**):**

First look at atom directly attached to the double bond;

(A)

2. If same atomic number, higher mass = higher priority (D > H, $^{13}C$ > $^{12}C$, etc.)
3. If the directly attached atoms are identical, move to the highest priority atom attached to the each of those atoms; the group substituted with higher atomic number substituent gets higher priority (basically apply rules 1 and 2 to next atom out, and so forth).

(B)

4. If a substituent is doubly or triply bonded to another atom, treat it as if it is two or three separate single bonds; **meaning:**

(C)

**Notes:**

# Easy EZ

**I.** The generic structures representing E- and Z- isomers should be used in deciding whether an alkene is the E- or Z- isomer:

**Note from the generic structures that you only need to know which substituent is high and which is low ON EACH CARBON.** The best way to do this is to start by dividing the alkene into two halves:

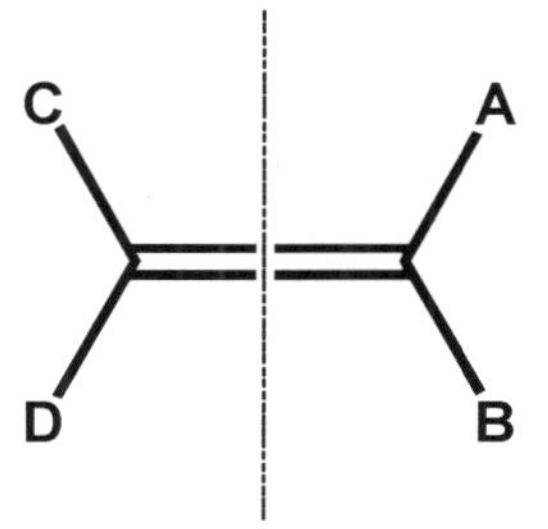


Your task is then to determine whether **A** or **B** is higher priority on the right-side carbon and whether **C** or **D** has higher priority on the left-side carbon. Once you do this, you simply match your assignments to one of the generic representations, i.e., if the higher priority substituent on the left side and right side carbons both point to the same side of the double bond (both up or both down) it is a Z- isomer, whereas if one of the high priority substituents is up and one is down (on opposite sides of the double bond) it is an E- isomer.

**II.** When we talk about "substituents" in the context of assigning priorities, we are talking about whatever is attached to the carbon atoms having the double bond in question; **these are NOT the same substituents that are listed in the name of the compound!** In the structure above, **A**, **B**, **C**, and **D** are the "substituents", regardless of their identity. For example, the "substituents" on the alkene below are outlined with dashed lines:

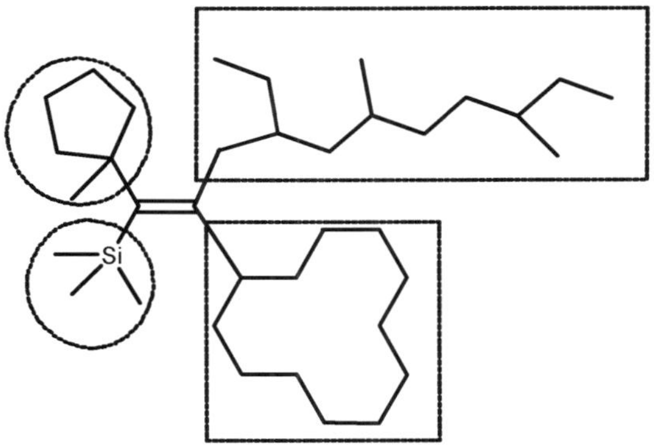

**III.** Now you must prioritize the substituents on each carbon as "high" or "low" using the priority rules.

**1.** first consider the atom directly bound to the double bond carbon. Rank them by atomic number - higher atomic number, higher priority:

I > Br > Cl > S > F > O > N > C > H

If the atomic number is the same but the mass is different (comparing isotopes of the same atom), the heavier isotope is higher priority. Thus: $^{13}C > ^{12}C > ^{2}H$ (denoted D, for deuterium) > $^{1}H$

**2.** a. If the directly bound atoms are the same (A and B in the figure), move out to the highest priority atom (use atomic number) bound to A and B (A1 and B1 in the figure). **NOTE:** you ONLY consider the next atom out if the first atoms (A and B) are the same; otherwise they do not matter at all!!!!!

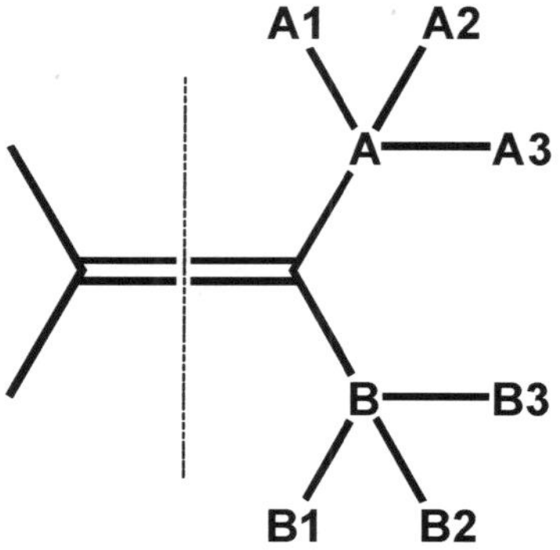


b. If A1 has higher priority than B1, then the "A" substituent has higher priority. If (and only if) A1 and B1 are ALSO tied (same atom), you have to move to the next-highest priority atom (A2 and B2) on A and B. If A2 is higher priority than B2, the "A" branch is the higher priority substituent. If A2 and B2 are tied, compare the next-highest priority substituents on A and B (A3 and B3). If A3 is higher than B3, the "A" branch is higher priority.

c. **If** A1, A2, and A3 match B1, B2, and B3, **then** you must begin comparing atoms attached to A1 and B1 using the same methodology described for comparing atoms on A and B, and continue moving farther from the initial attachment point in this manner until you find a difference between the A and B branches that allows you to make an assignment.

Examples:

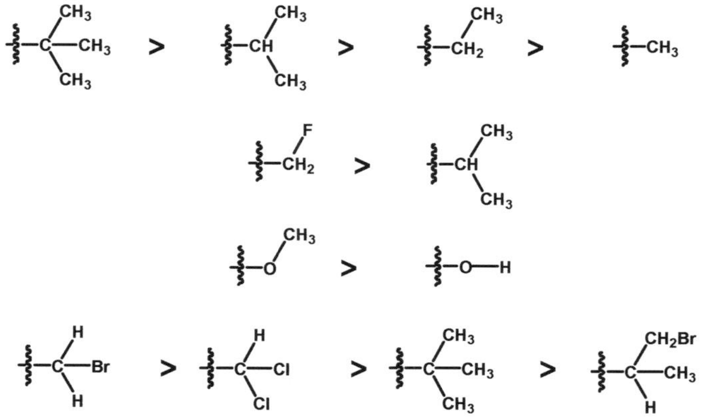

**3.** Multiple bonds are "broken open" to form the same number of single bonds to the same atoms. This method is sometimes called the "break and duplicate" approach; break the pi bonds, duplicate the atoms to which they were bound on each side of where the pi bond was:

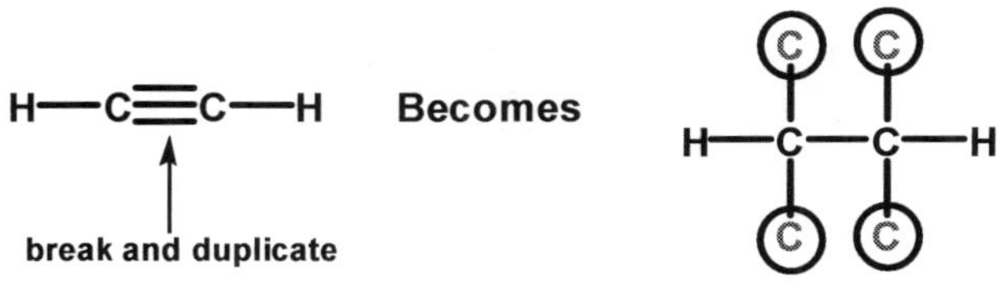

**NOTE:**
**the circled atoms do not have any other atoms attached! They are imaginary constructs that serve only to help us prioritize, and have no other chemical significance!**

# Examples:

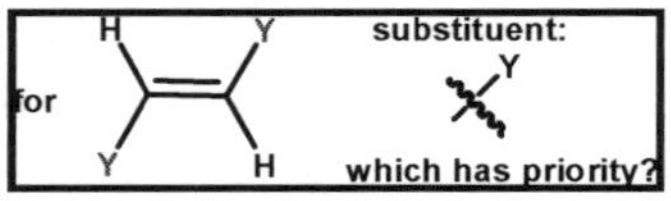


**Rules 1-3 let us rank these:**

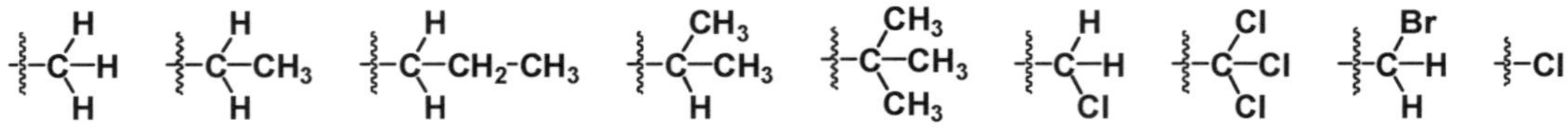


**Lowest priority** **Highest priority**

**Rule 4 lets us deal with multiple bonds in substituents:**

treat 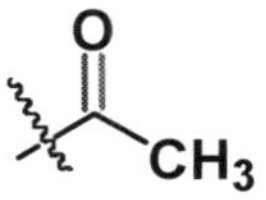 as

treat 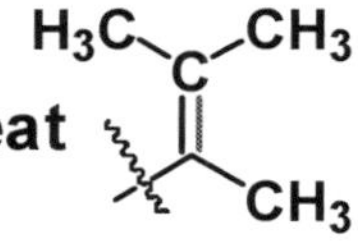 as

treat 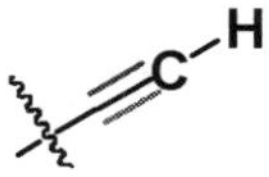 as

**Highest priority**

**Lowest priority**

# Notes:

# Priority Rules Problem

Example. Designate each double bond as *E*- or *Z*-.

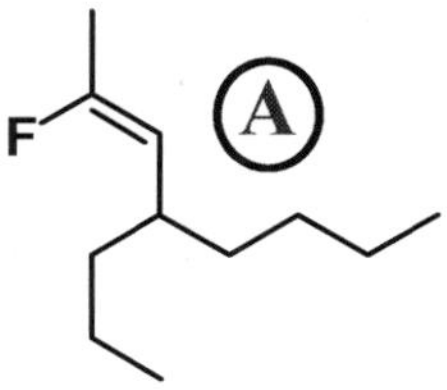

(A)

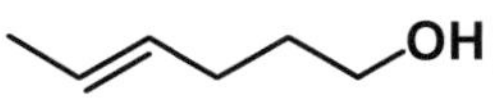

(B)

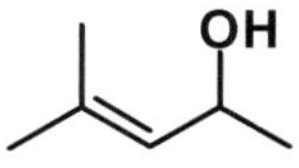

(C)

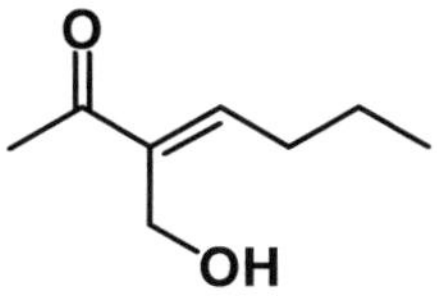

(D)

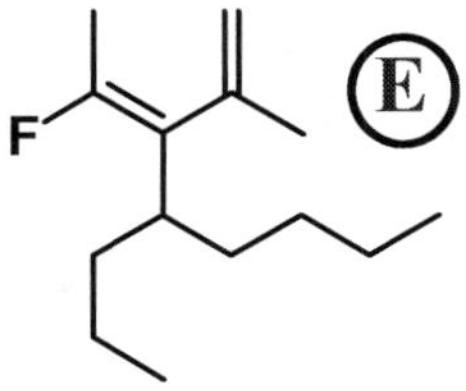

(E)

**Notes:**

# Properties of Alkenes: What We Know

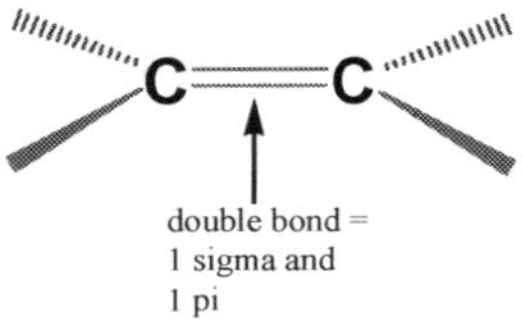


(A)

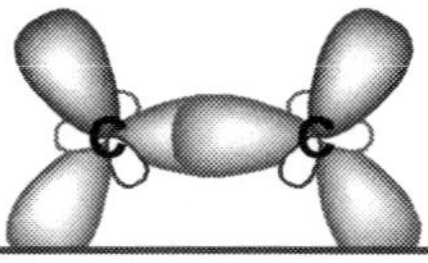


(B)

(C)

**Notes:**

# Observations:

Now that we can name alkenes, let us turn to some structure-property relationships.

1.

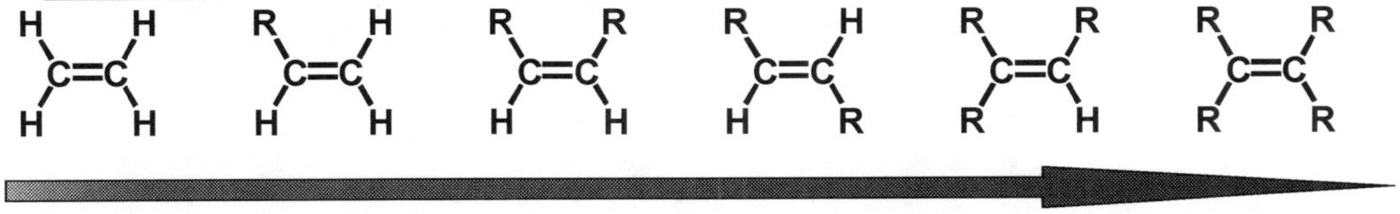


**LESS STABLE** **MORE STABLE**

2.

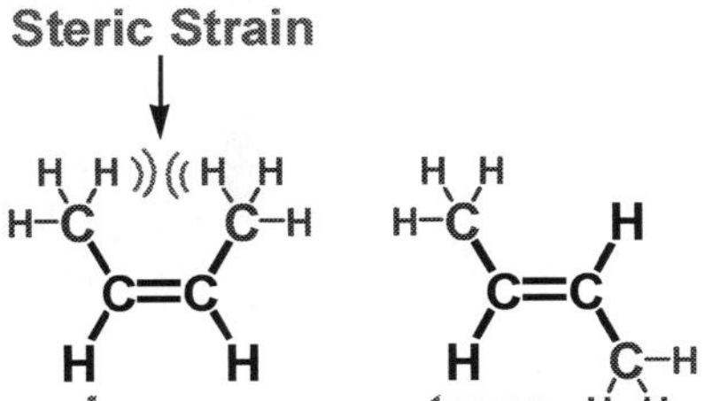


## Notes:

# Measuring Alkene Stability

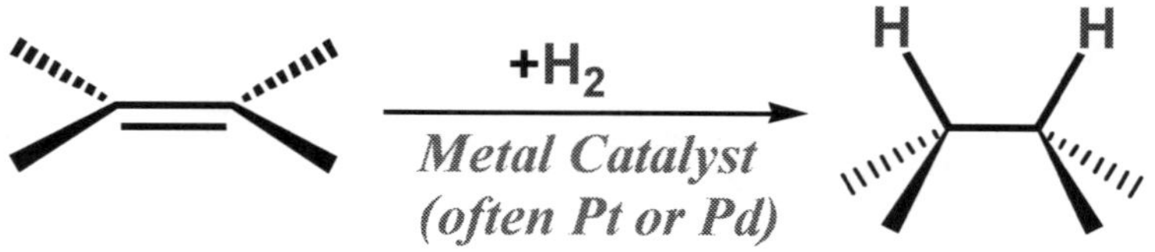


The Heat of Hydrogenation tells us about: (A) ____________________

A more stable alkene has a (B) ____________________

Higher energy!! (more heat given off) ↑ … Lower energy!! (Less Heat Given Off)

| Structure | heat of hydrog. (kcal/mol) |
|---|---|
|  | 30.1 |
| 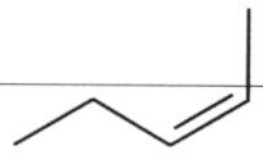 | 29.1 |
| 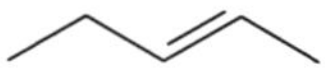 | 28.2 |
| 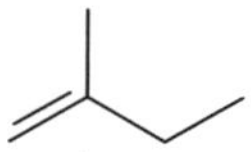 | 27.8 |
| 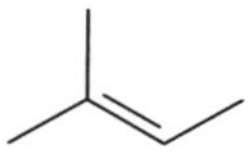 | 26.7 |

Less Stable ↓ More Stable

## Notes:

# Properties: more $e^-$ density between $sp^2$ C's

alkenes have a region of higher electron density at the pi bond and may have a dipole moment, **Because**

Ⓐ

and

Ⓑ

Ⓒ

Ⓓ

**Notes:**

# Preparation of Alkenes from Alcohols

A common way to prepare alkenes is via an **elimination reaction**. A general representation of such a reaction is given below:

(A)

This reaction is particularly favorable when a stable (or trappable/removable) species XY is released. If we use an alcohol as our starting material, we can eliminate the OH (hydroxy) group with a proton to give water:

(B)

Since water is removed, this reaction is referred to as a

(C)

**Notes:**

# Mechanism of Dehydration: E1

These are examples of **E1** reactions (E for elimination, 1 for unimodular)

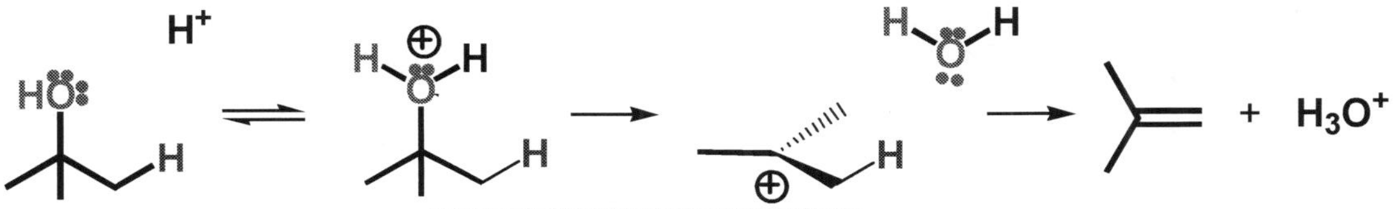


A

The **E1** reaction proceeds via carbocation intermediate formation as the rate limiting step.

B

**Notes:**

## Ⓐ Observation:

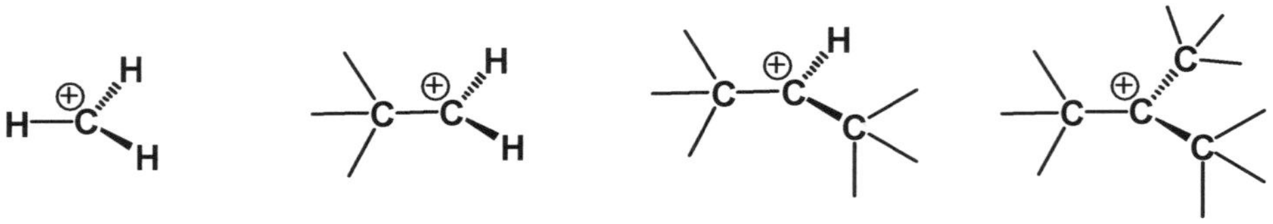


methyl　　$1^o$　　$2^o$　　$3^o$

STABILITY

LESS STABLE　　MORE STABLE

## Ⓑ Explanation:

## Ⓒ Ramification:

REACTIVITY

$H_3C\text{-}OH < 1^o < 2^o < 3^o$

**Notes:**

# Stability of Carbocations

Compare cation stability trends with:

1. Anion stability (acidity) in LS 1

2. Radical stability, LS 2

There are no such stabilizing interactions in the methyl carbocation or radical!

**Notes:**

# Regioselectivity: Which H is Eliminated in E1?

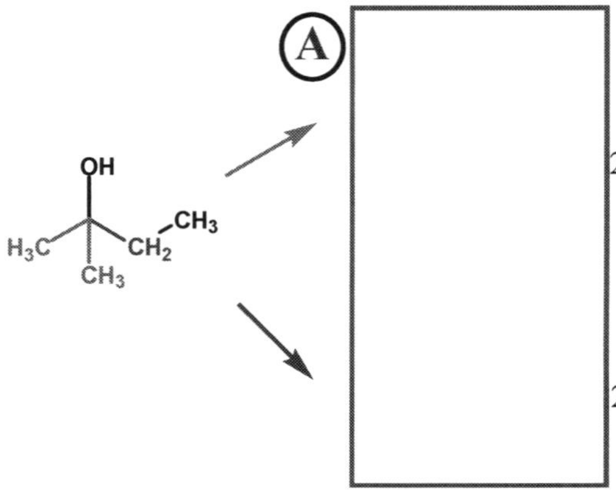


2-methyl-1-butene

2-methyl-2-butene

Two regioisomers as possible products

The mechanism tells us that they both arise from the same intermediate carbocation:

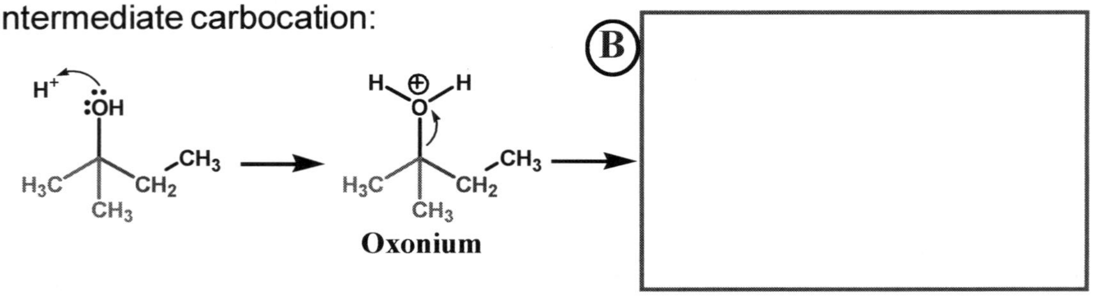


**Notes:**

# Reaction Coordinate Diagrams

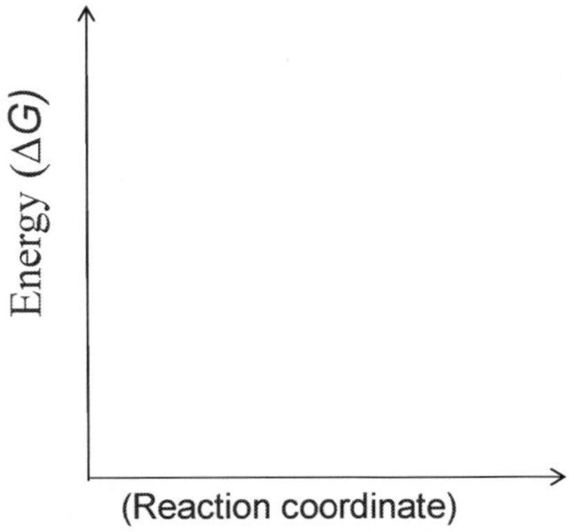


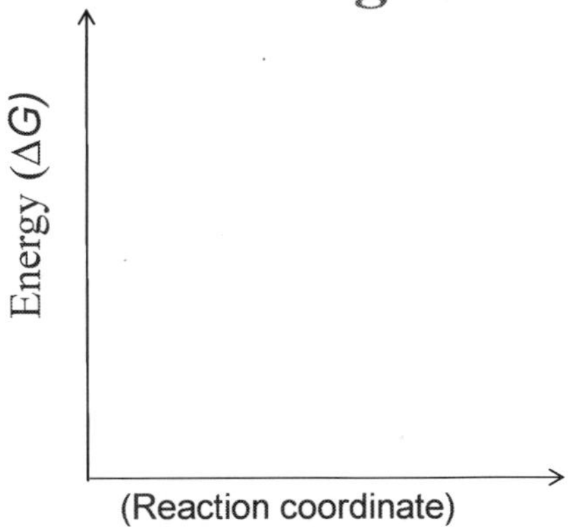


| | |
|---|---|
| Number of Steps | (A) |
| Energy of Activation | (B) |
| Height on Energy Axis | (C) |

**Notes:**

# Regioselectivity

**Observation = "Zaitsev's Rule:**

(A)

**Explanation = Reaction coordinate diagram:**

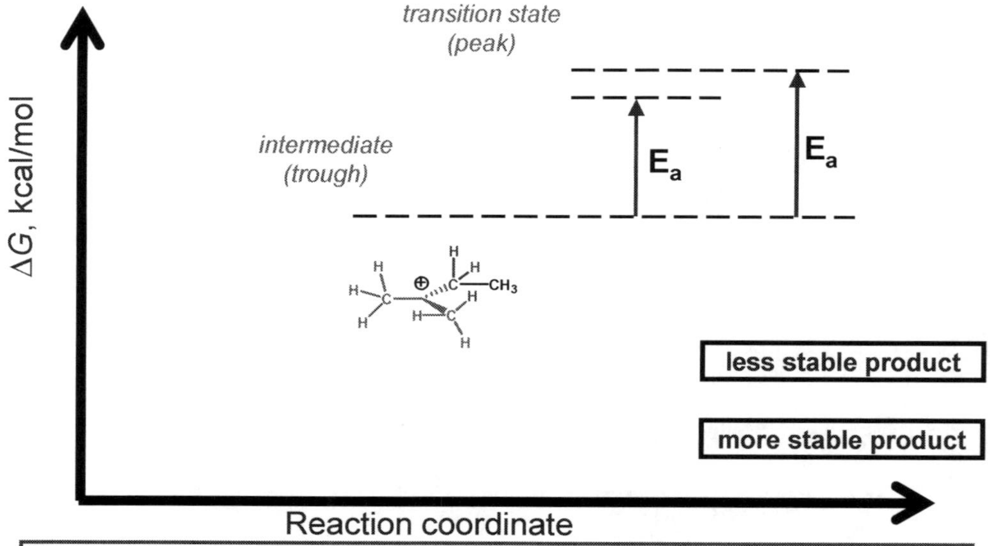


(B)
Remember

**Notes:**

# Zaitsev's Rule in Action

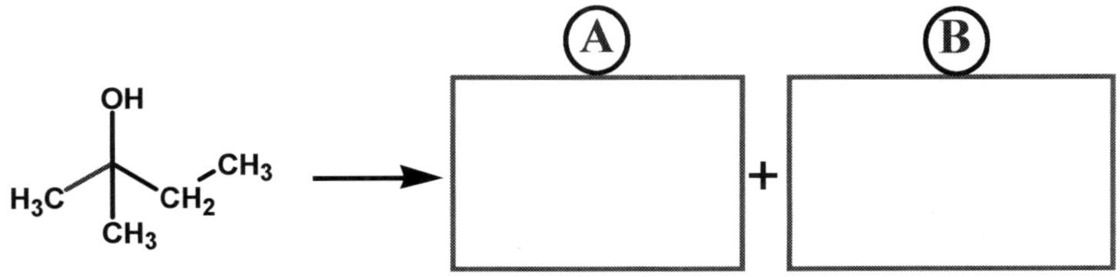


90% **MORE STABLE** | 10% **LESS STABLE**

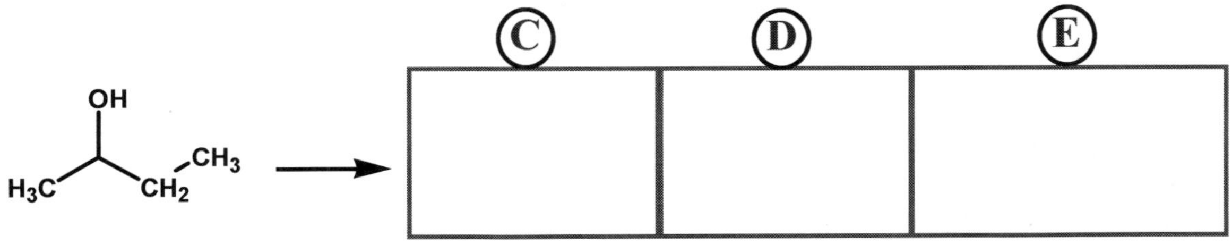


| C | D | E |
|---|---|---|
| 56% **MOST STABLE** *trans-* disubstituted | 32% *cis-* disubstituted | 12% **LEAST STABLE** monosubstituted |

**Notes:**

# Synthetic Problem

Example. Give the possible elimination products and indicate which is expected to be the major isomer.

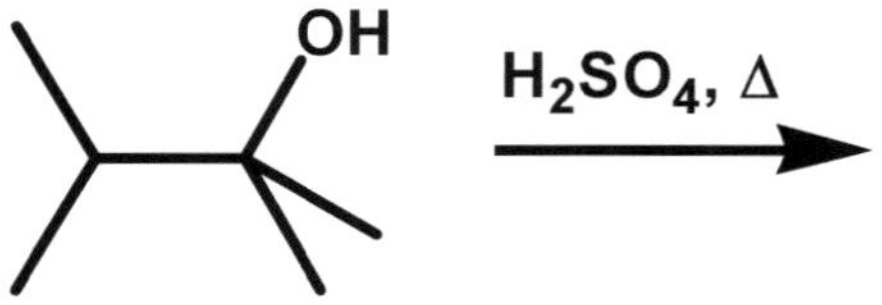


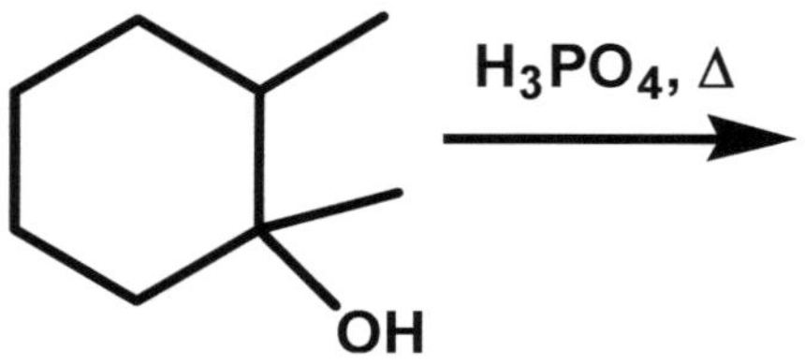


**Notes:**

# Rearrangement of Carbocations

Reactivity trends depend upon the enrgy of interactions, intermediates, transition states, etc. Reactions that result in a lower energy species will be favorable. High energy intermediates such as carbocations can undergo intramolecular rearrangement to a lower energy structure.

**Molecules will always**

(A)

Example: a hypothetical primary carbocation adjacent to a tertiary carbon can rearrange to make the more stable tertiary carbocation by moving a hydrogen (B)

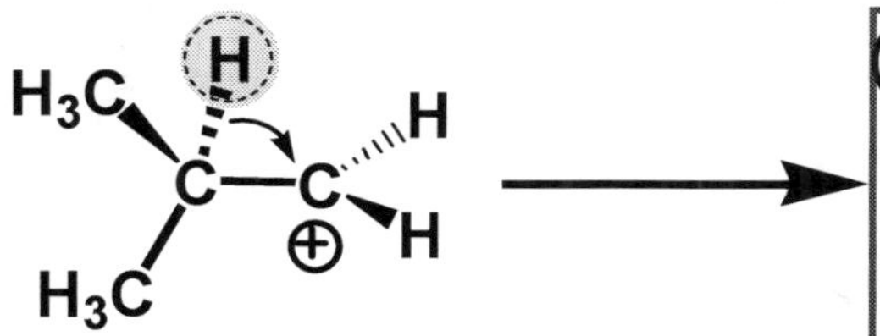


Primary Carbocation

(C)

**Notes:**

# Rearrangement of Carbocations

Alkyl groups can also migrate to create a lower energy species. Note that neither a hydride nor an alkyl group will undergo a 1,2-shift to form a HIGHER energy species.

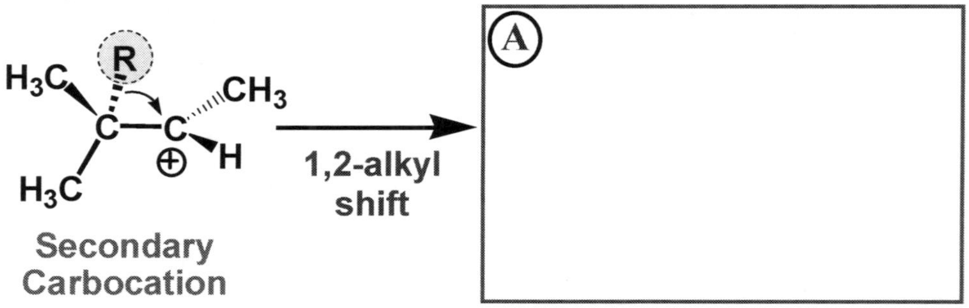


If a carbocation formed as an intermediate in an elimination reaction occurs, we will observe different products resultant from the new, more stable carbocation.

Rearrangement to a more stable carbocation is very fast and

(B)

**Notes:**

# Rearrangement of Carbocations

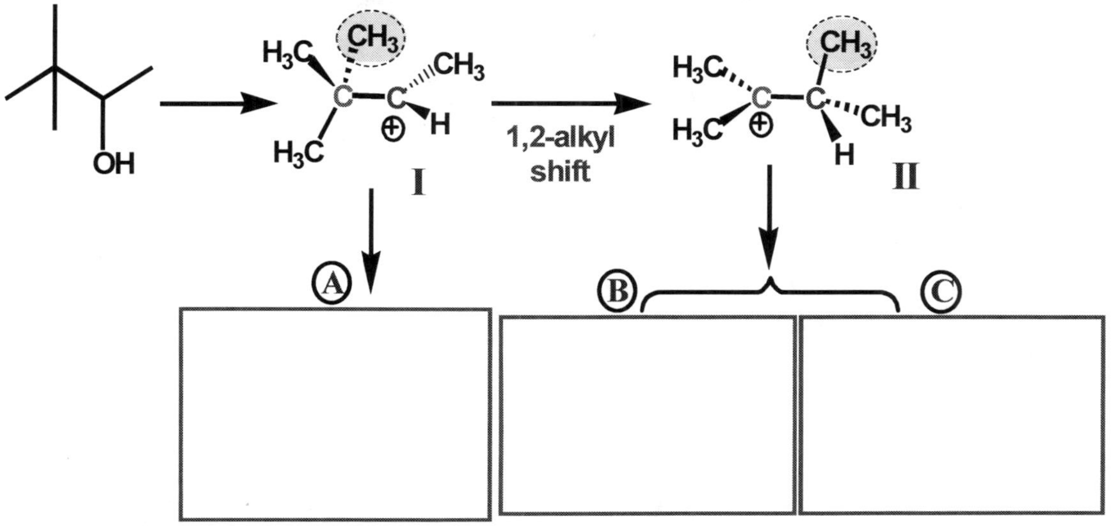


(D) **hints:**

**Notes:**

With cycloalkanes, the ring size can change when a 1,2-shift occurs, and rearrangement can be driven not only by stabilization of the cation, but also

(A) ____________________

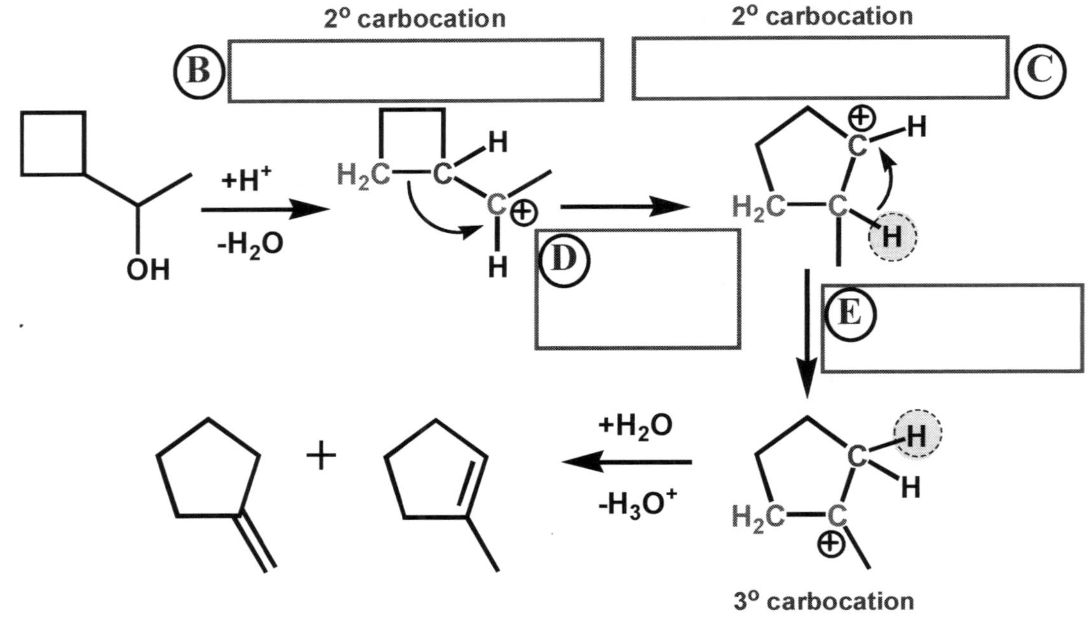


**Notes:**

# Preparation of Alkenes from Alkyl Halides

Alkenes can also be prepared from alkyl ahlides via the elimination of HX, a process known as (A)

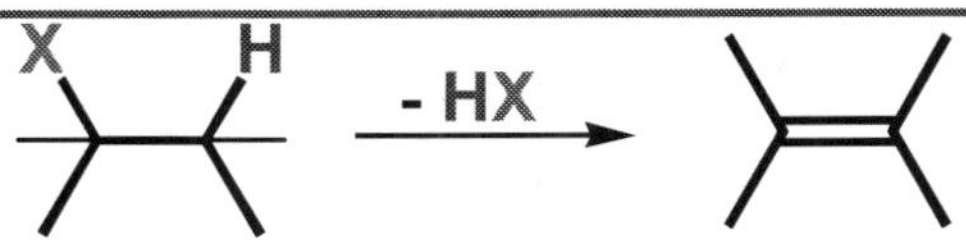


X = Cl, Br, I

These reactions are elimination reactions, and we will discuss two different mechanisms. The more stable (Zaitsev) alkene will be formed when possible, just as we observed for the preparation of alkenes from alcohols.

**HX is a strong acid;** generally, rxn to give strong acid is unfavorable! Thus:

(B)

(C)

**Notes:**

# The E2 Reaction

The mechanism, known as E2, is fairly straightforward. The base extracts a proton ***in concert*** with elimination of a halide leaving group. Such a single-step mechanism is known as (A)

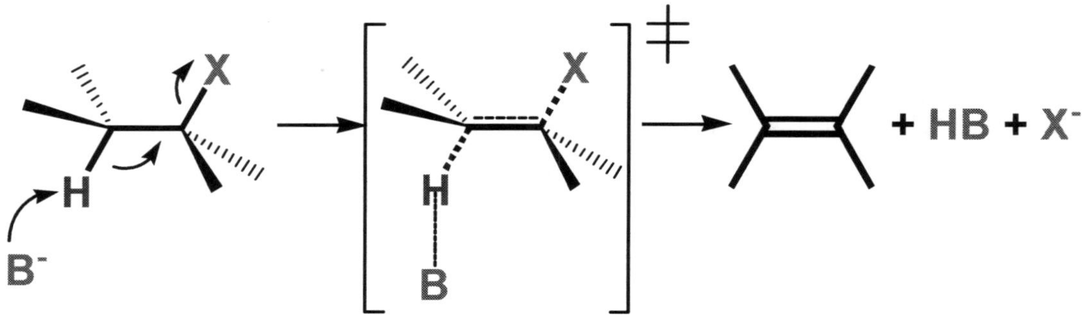


Because E2 occurs via a **concerted mechanism,**

(B)

**Notes:**

# The E2 Reaction: reactivity rate

E2 is a bimolecular (thus the 2) elimination (thus the E) reaction that can be used to prepare alkenes. Because the reaction is bimolecular, the rate depends on the concentration of two reagents. Here,

(A)

The rate limiting step involves breaking the R-X bond;

(B)

RF < RCl < RBr < RI

→

**REACTIVITY**

The rate limiting step involves R-X bond cleavage.

**Notes:**

# The E2 Reaction

A reaction coordinate diagram helps predict the product formed when more than one choice is available.

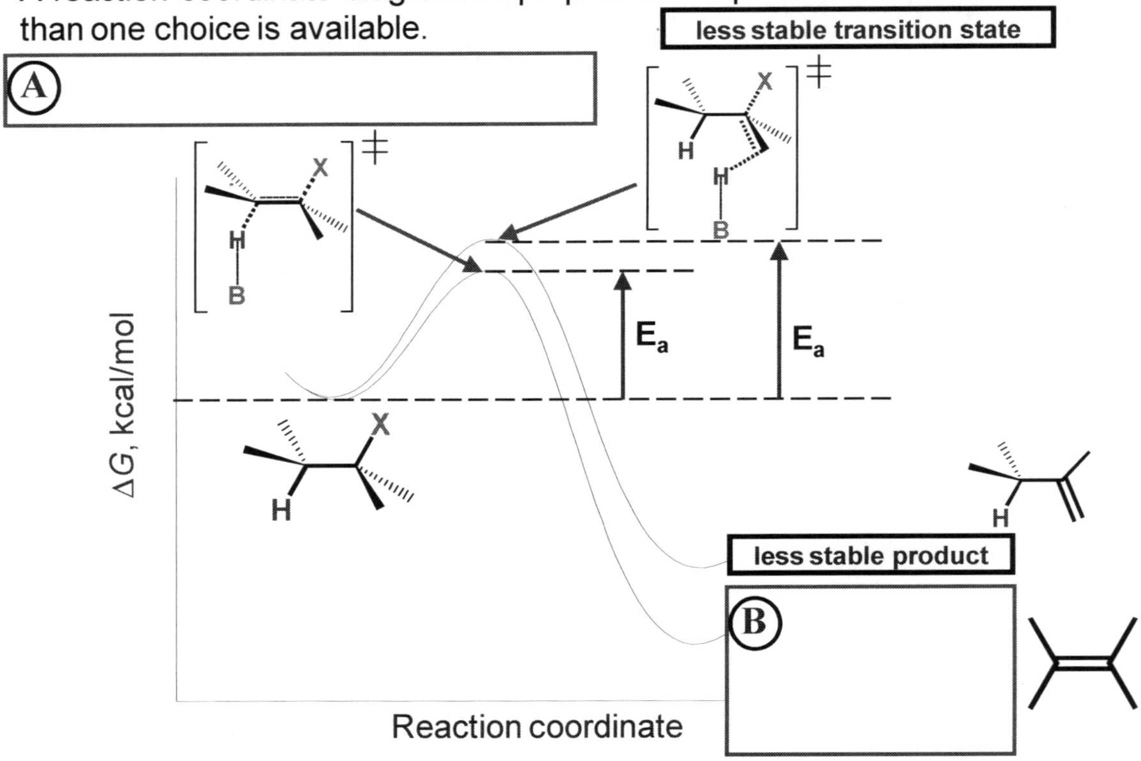


**Notes:**

# Stereoelectronics in the E2 Reaction

In order for the E2 reaction to work, the H and X to be eliminated must be **antiperiplanar**.

"Antiperiplanar" means the two groups are in the *anti-* conformation (see alkane conformation discussion in LS 2) and in the same plane.
This is what is known as a **stereoelectronic** effect,

(A)

**Notes:**

The stereoelectronic requirements of the E2 reaction must be considered when we attempt to predict products of a reaction:

**Explanation**

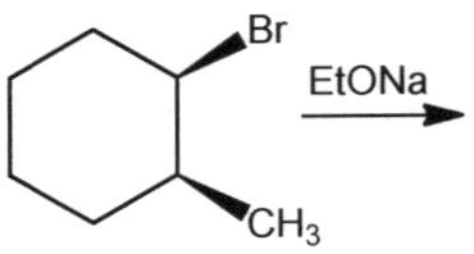


A

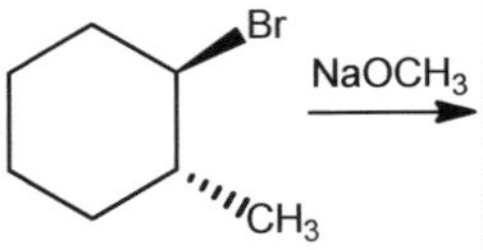


B

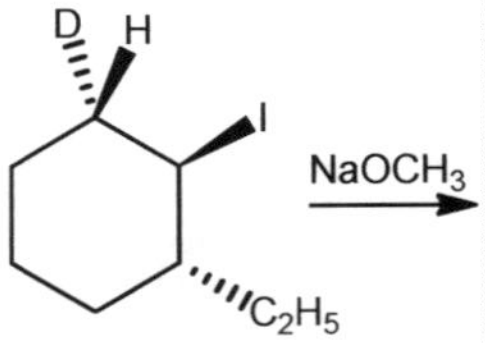


C

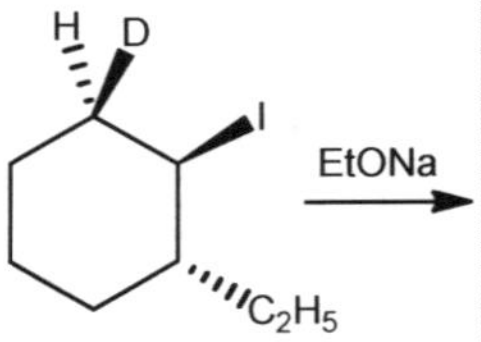


D

**Notes:**

# E2 and the Stereoelectronic Effect

Particular care must be taken when examining eliminations from cyclohexane. In order to gauge geometry carefully, we need to draw it in the chair conformation. Once we do this, it is easy to see that in order to have an antiperiplanar relationship with a hydrogen, the halogen must be in an axial position:

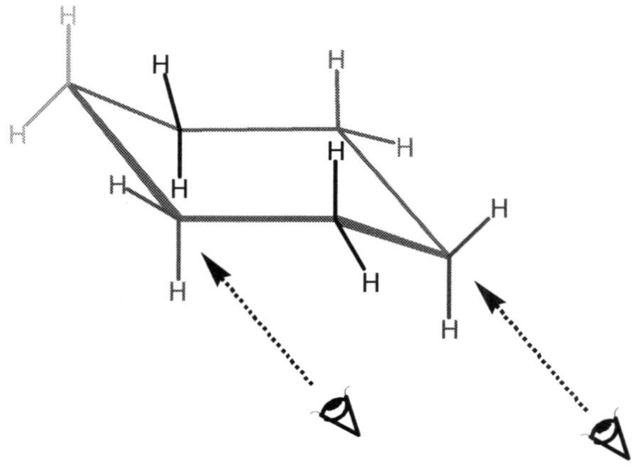

All staggered.

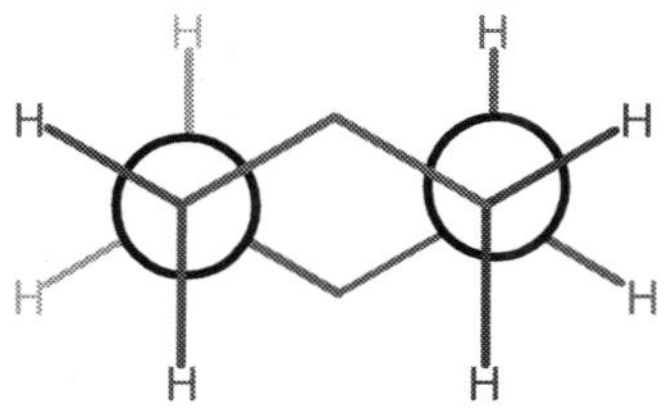

"Antiperiplanar" means the two groups are in the *anti*-conformation with respect to one another!

buildamodelbuildamodelbuildamodelbuildamodelbuildamodelbuildamodelbuildamodelbuildamodelbuildamodelbuildamodelbuildamodelbuildamodelbuildamodelbuildamodel

**Notes:**

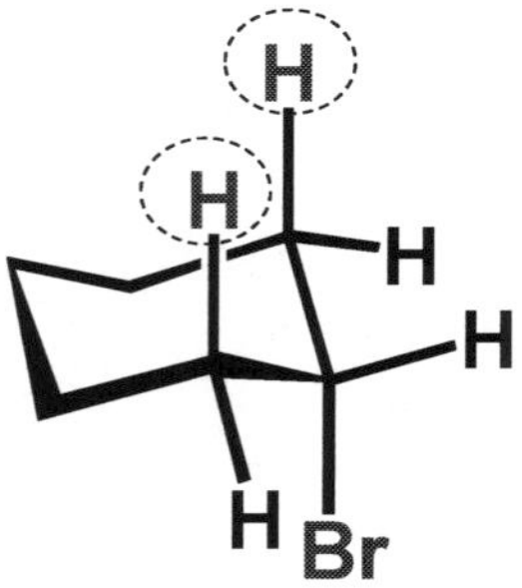


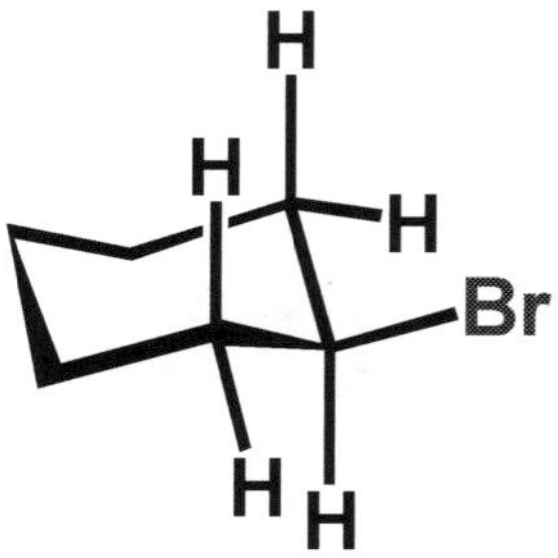


(A)

(B)

buildamodelbuildamodelbuildamodelbuildamodelbuildamodelbuildamodelbuildamodelbuildamodelbuildamodelbuildamodelbuildamodelbuildamodelbuildamodelbuildamodel

**Notes:**

Example. The *cis*- and *trans*- isomers of 1-bromo-4-*t*-butylcyclohexane are shown. The *cis*- isomer reacts 500 times faster than with the *trans*- isomer. Use your knowledge of stereoelectronic requirements for the E2 mechanism and the stability of cyclohexane conformations to rationalize this observation.

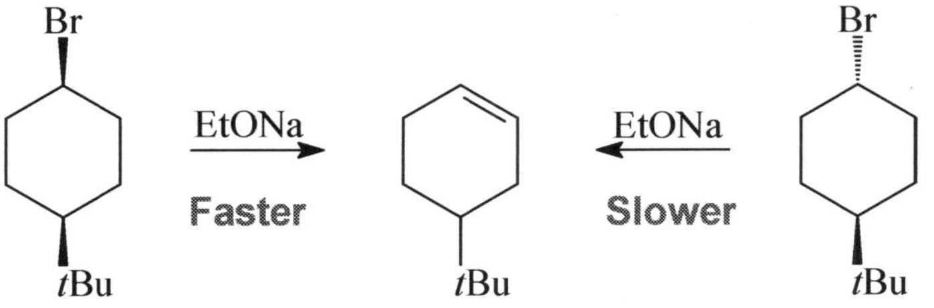


Answer:

(A)

(B)

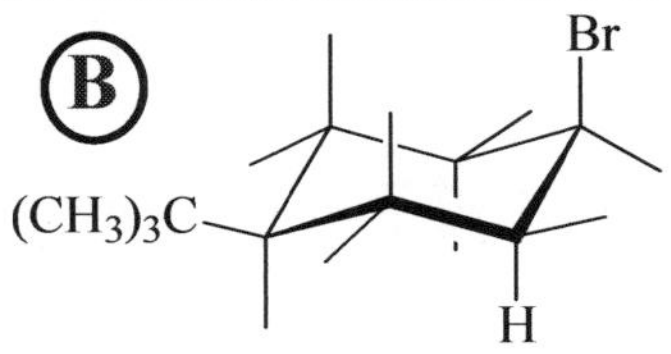


(C)

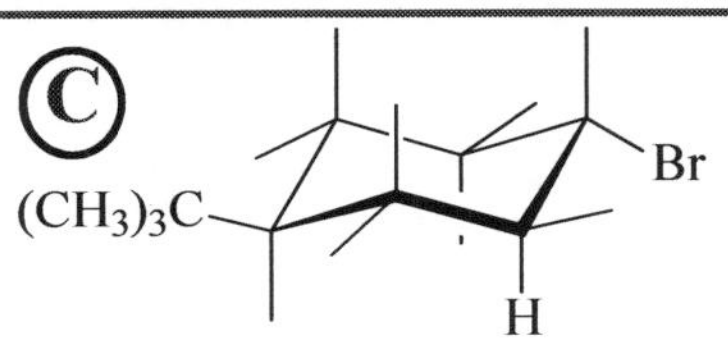


**Notes:**

# The E1 Reaction (preview)

Alkenes can also be prepared from alkyl halides via the **E1 reaction**, in the **absence of base**:

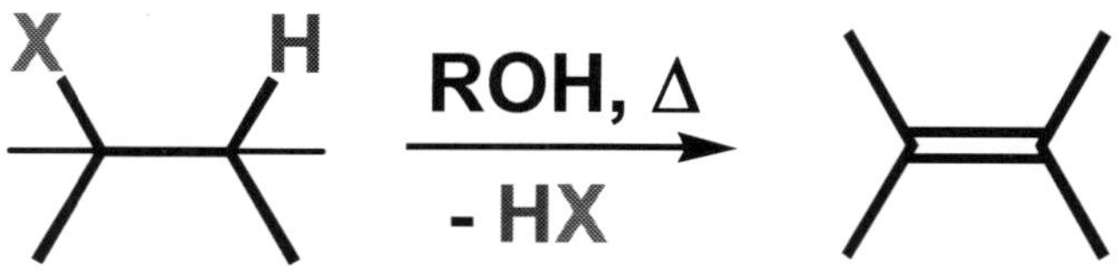


The trend in reactivity depending on whether the starting material is a primary, secondary, or tertiary alkyl halide, as was noted for dehydration of alcohols by the E1 mechanism, and also on the R-X bond strength (R-X bond has to break to form the carbocation):

RF < RCl < RBr < RI → **REACTIVITY**

$H_3C$-X < 1° < 2° < 3° → **REACTIVITY**

We saw the E1 mechanism for dehydration, so this mechanism should be very familiar...

**Notes:**

# The E1 Reaction

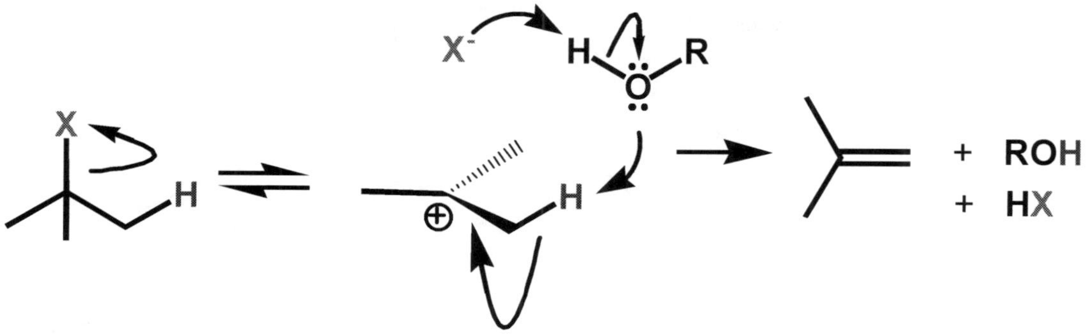


Formation of the carbocation is (A) [          ]

Of course, the cation can

(B) [          ], leading to multiple potential products.

(C) [          ] applies to determining the anticipated product.

**Notes:**

# Reaction Examples

Example. Give the predominant product of each reaction. Underline the major product in each case. Label each reaction as proceeding via an E1 or E2 mechanism; note any instances in which carbocation rearrangement occurs.

(A) 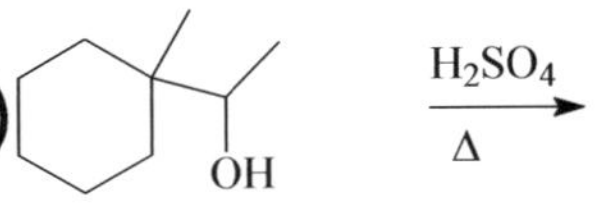


(B) 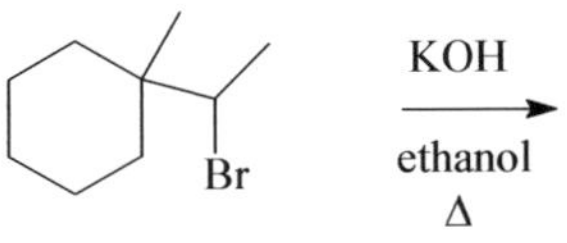


(C) 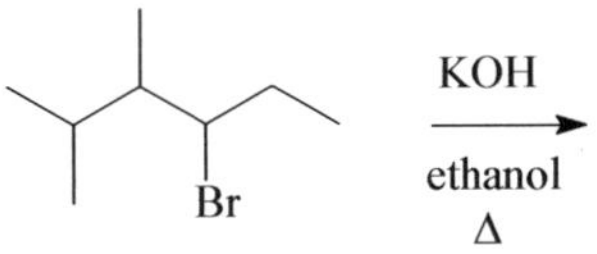


(D) 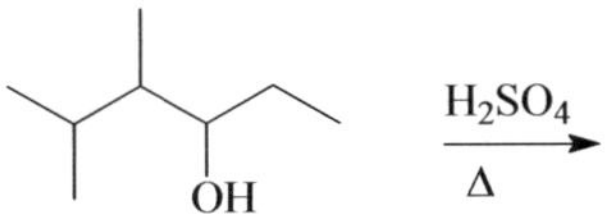


**Notes:**

# Lecture Set 5:
# Reactions of Alkenes

**Suggested Reading:**

**Suggested Problems:**

# Addition Reactions of Alkenes

Alkenes can be prepared by elimination reactions. Alkenes can also undergo

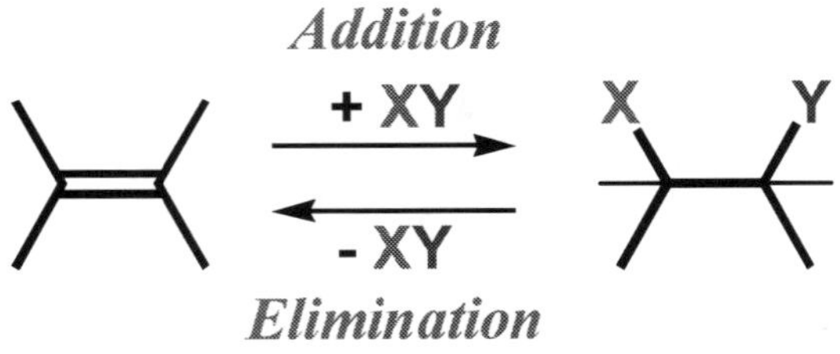


(A)

to produce other useful products:

A stereochemical/mechanistic issue to be addressed is whether a given addition reaction occurs with the resultant substituents adding *syn-* or *anti-* to one another:

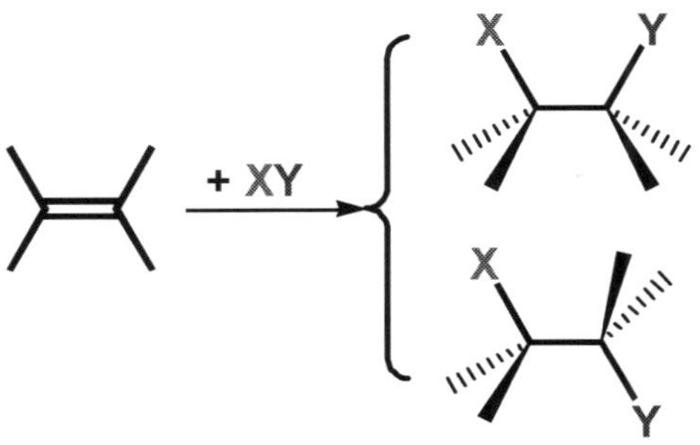


(B)

(C)

**Notes:**

# Reactions of Alkenes: Hydrogenation

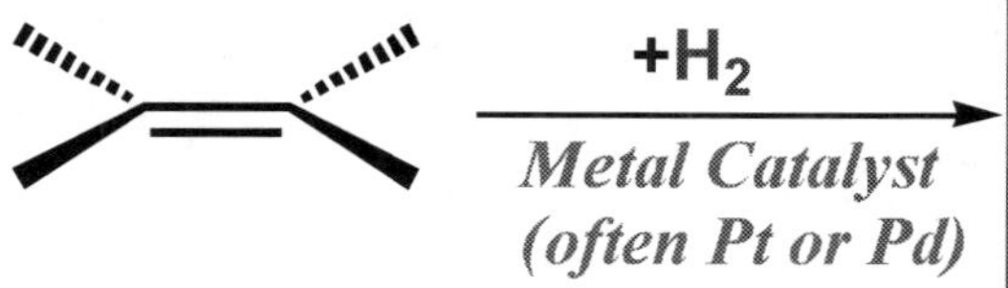


A

This type of catalytic hydrogenation leads to *syn* addition:

B

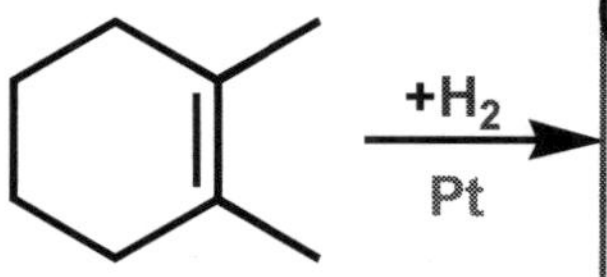


C

Syn-addition of $H_2$ to cycloalkenes can lead to *cis*-disubstituted cycloalkanes.

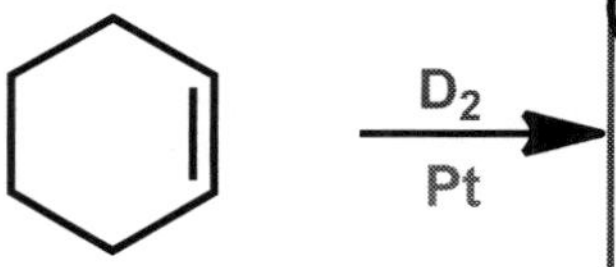


D

We can use the deuterium isotope of hydrogen to further confirm the *syn*-addition

**Notes:**

# Reactions of Alkenes: Electrophilic Addition

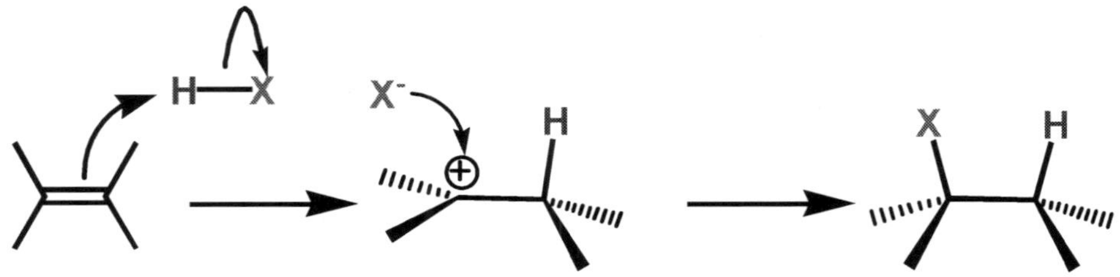


This addition reaction is called (A)

The mechanism is essentially the reverse of

(B)

Once again, the rate limiting step is

(C)

On the basis of acid strengths and carbocation stabilities we can rationalize trends in alkene reactivity and observed product distributions …

**Notes:**

# Electrophilic Addition: Reactivity Trends

The observed rate of reactivity for the acid (HX):

HCl < HBr < HI

**REACTIVITY**
**Acidity**
**Nucleophile Strength**

Consider the possible products of this reaction:

*2-iodo-2-methylpropane* *1-iodo-2-methylpropane*

**Notes:**

# Hydrohalogenation: Products

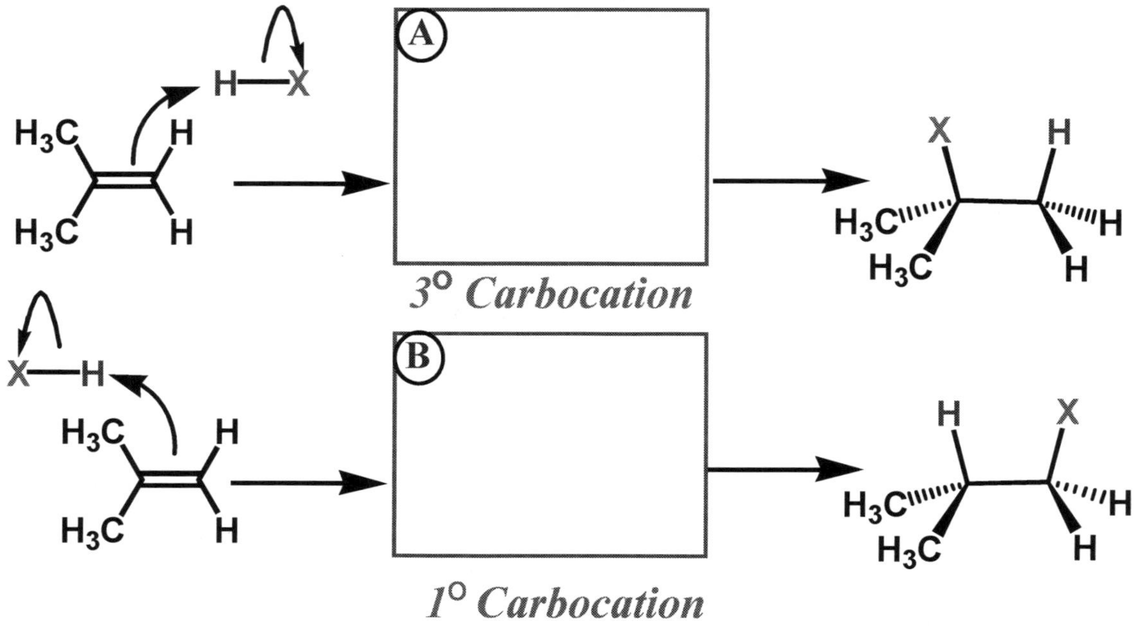


The major product is that derived from the more stable carbocation, in this case the tertiary cation.

**Notes:**

OBSERVATION= Markovnikov's rule:

(A)

EXPLANATION=

(B)

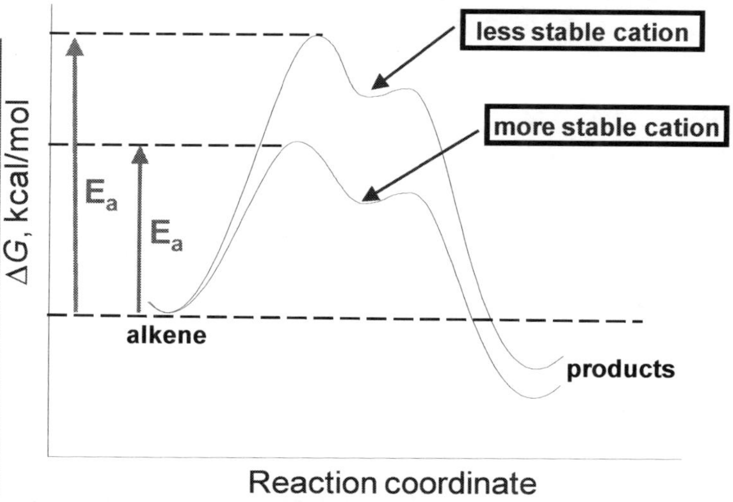


We can use the same type of energy-based arguments to rationalize Markovnikov's rule as we did to rationalize Saitsev's rule.

**Notes:**

# Electrophilic Addition: Problem

Example. Give the major products.

(A) 

(B) 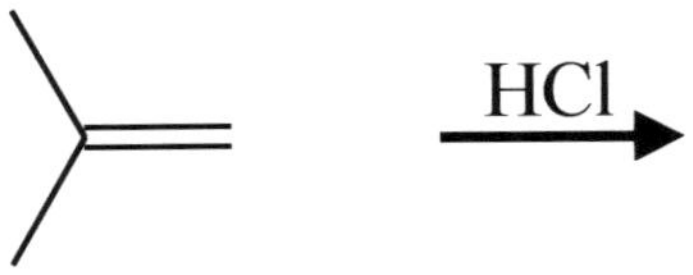


(C) 

(D) 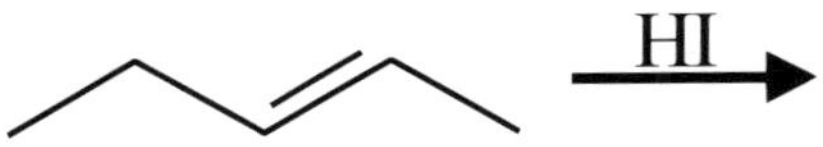


**Notes:**

# Watch for Carbocation Rearrangement

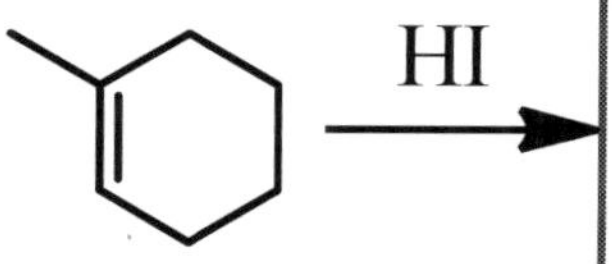


(A)

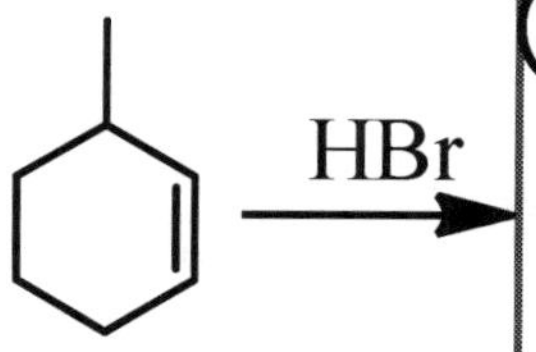


(B)

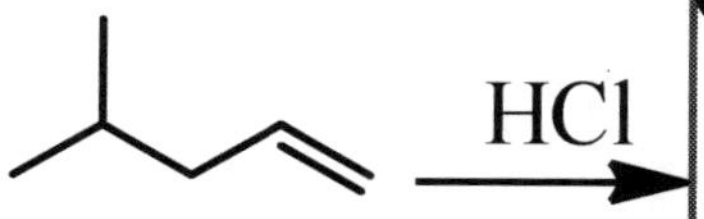


(C)

**Notes:**

# Preparation of Alcohols: Hydration

Addition of HX to an alkene is the reverse of dehydrohalogenation. We can also do a hydration reaction (addition of water to an alkene), which is the reverse of the dehydration of alcohols that we saw in dehydration (E1 route).

(A)

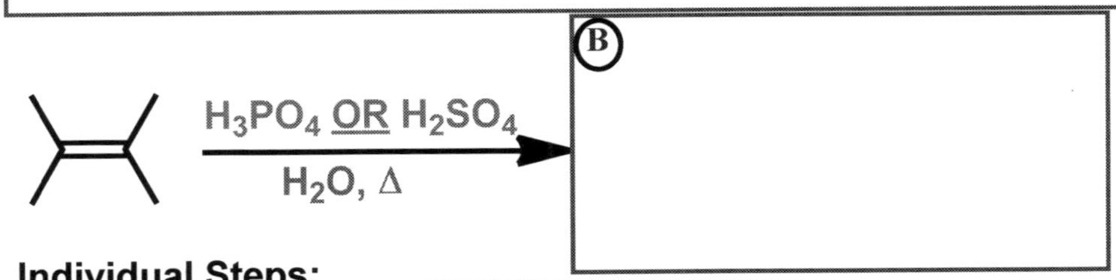


**Individual Steps:**

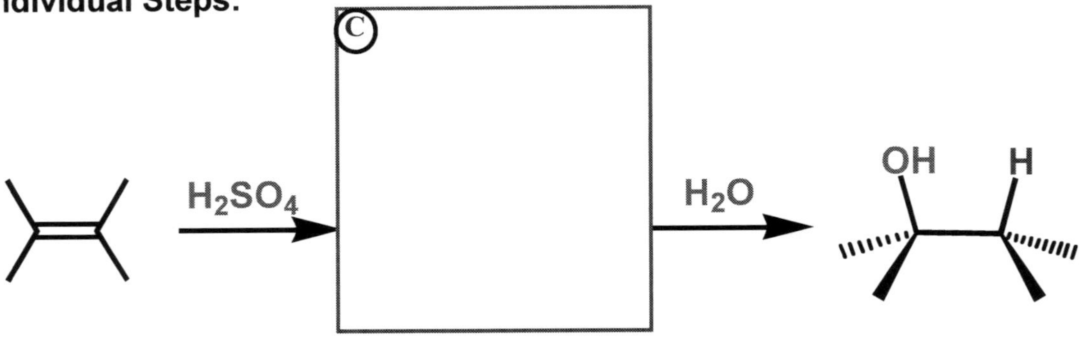


**Notes:**

# Alternative to Cations: Radicals!

Peroxide-initiated hydrohalogenation is an alternative to the E1 route:

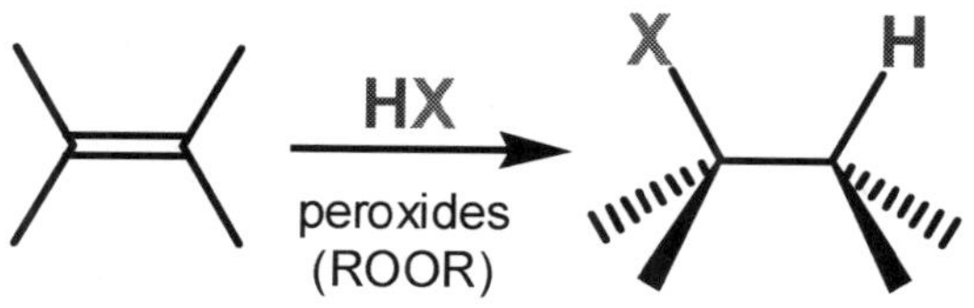


Ⓐ **Initiation**

Ⓑ **Propagation**

Ⓒ **Net Reaction**

**Notes:**

Termination occurs when any two radicals come together.

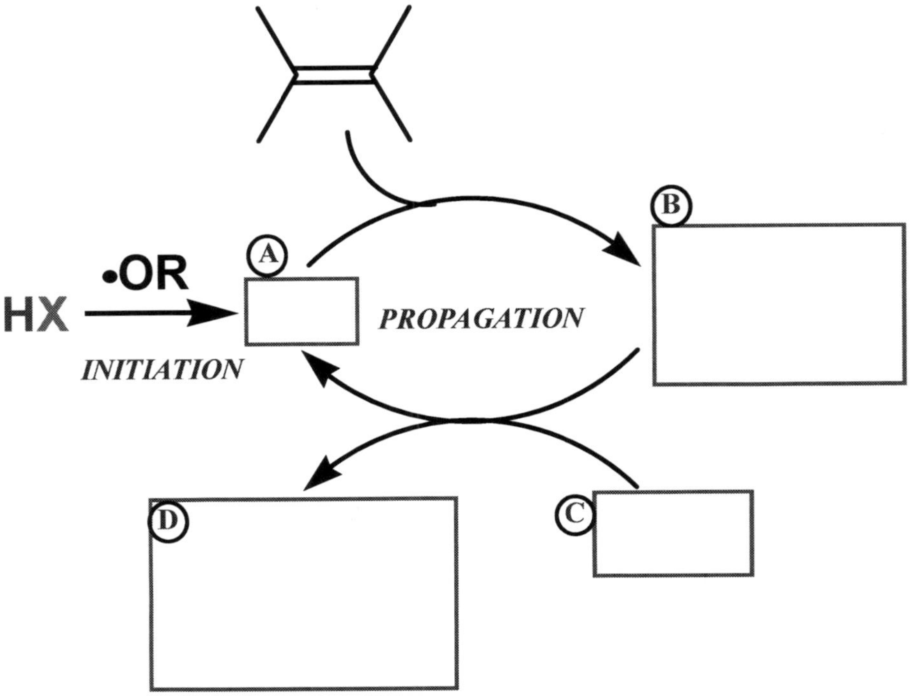

**Notes:**

# More Stable Radical Determines Major Product

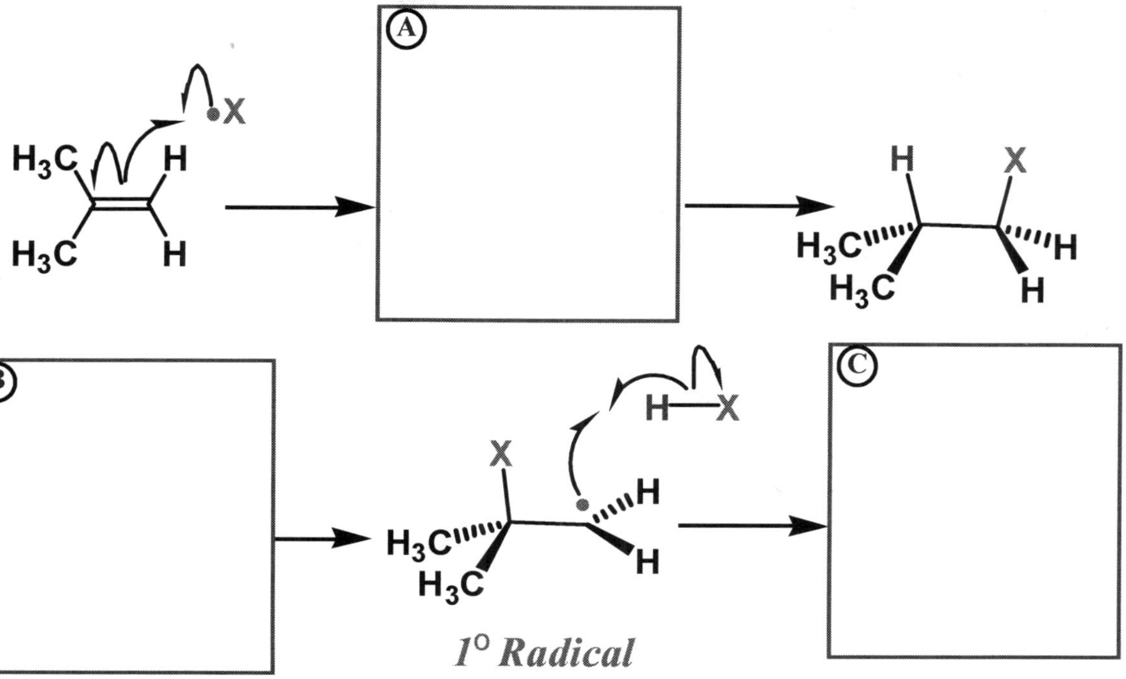


Once again, the major product is that derived from the more stable intermediate, in this case the tertiary radical.

**Notes:**

# Radical Reactions Review

**1.** Rank the four indicated C–H bonds from lowest (1) to highest (4) bond dissociation enthalpy.

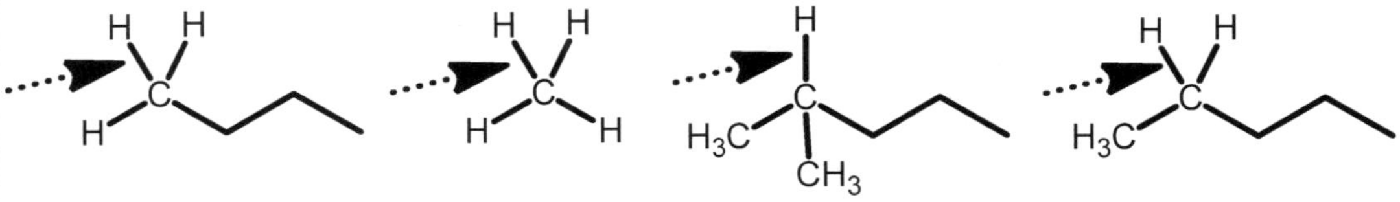


**2.** What is the major product of each reaction:

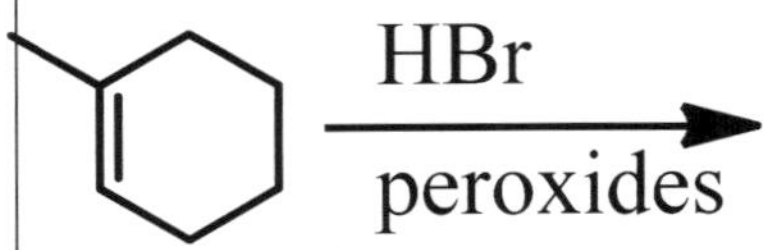


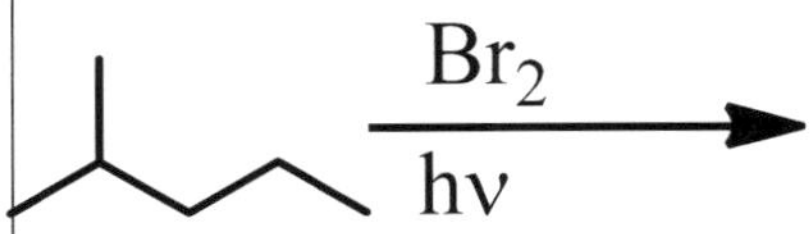


**3.** Draw the initiation step of peroxide-initiated hydrobromination of propene:

**4.** Draw the propagation step of peroxide-initiated hydrobromination of propene:

**5.** Draw the major product of the peroxide-initiated hydrobromination of 1-butene:

# Different Products for Different Conditions

Note that the peroxide-mediated (radical intermediate) hydrohalogenation yields the so-called "anti-Markovnikov" product, while the peroxide-free E1 hydrohalogenation (carbocation intermediate) yields the Markivnikov product. This gives a researcher access to whichever product is desired:

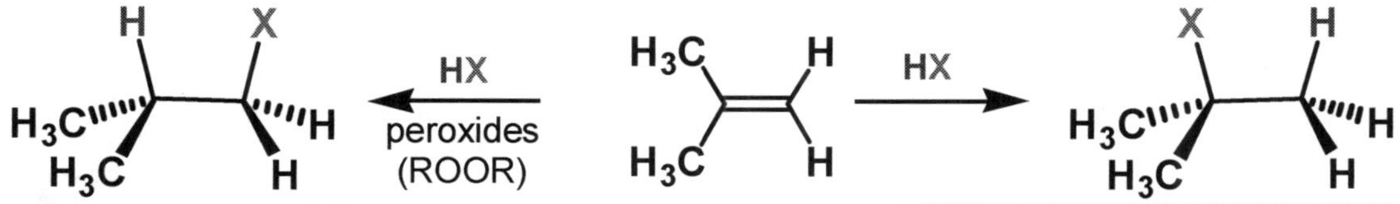


(A)

(B)

(C)

Remember the mechanism and you will not need to memorize this!

**Notes:**

# Halogenation of Alkenes

Halogens ($X_2$) can add to alkenes to form 1,2-dihalides (vicinal halides):

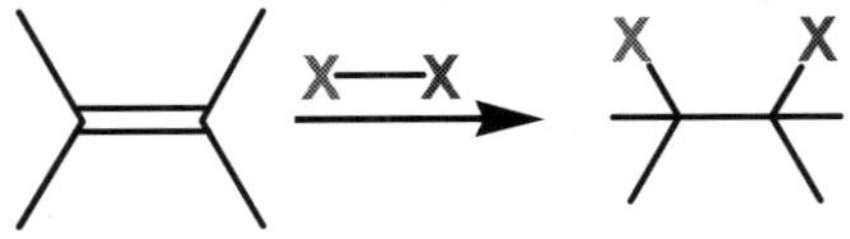


The mechanism shows us that the two halogen atoms add anti to one another (so if a cycloalkene is used, we will get a trans-1,2-dihalide).

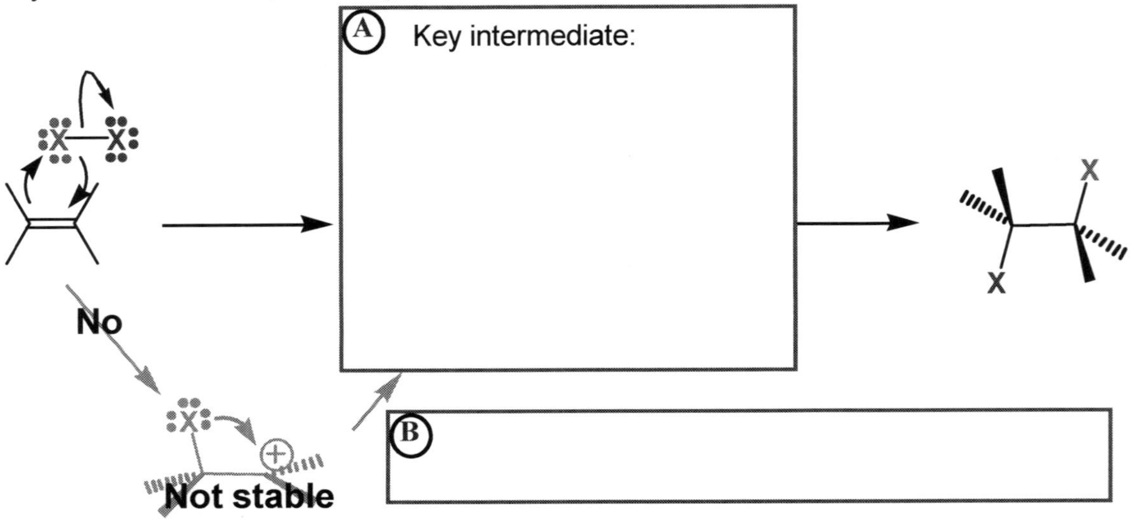


Formation of halonium is rate limiting. What about regioselectivity?

**Notes:**

# Halogenation

(A)

One can also use Br-Cl or I-Cl to make mixed alkyl halide (see example).

(B)

**REACTIVITY**

Example. Based on your knowledge of halogenation mechanism and electronegativity, draw the arrow-pushing mechanism leading to the major product of the reaction.

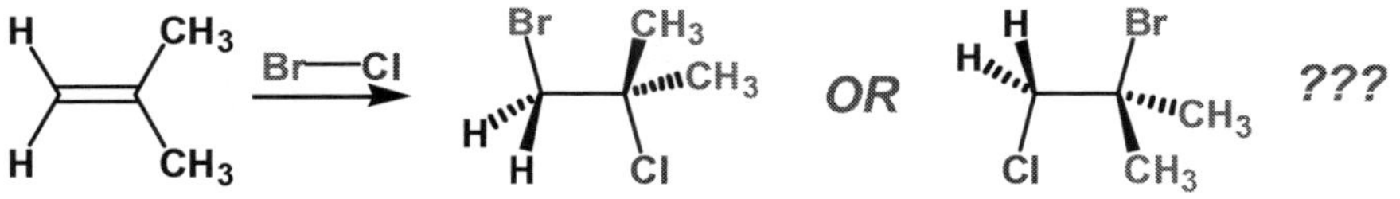


**Notes:**

**First:** will a chloronium or bromonium ion form?

The more electrophilic site of the polar X-X bond will attack the pi bond.

(A)

**Second:** to which carbon will the chloride add?

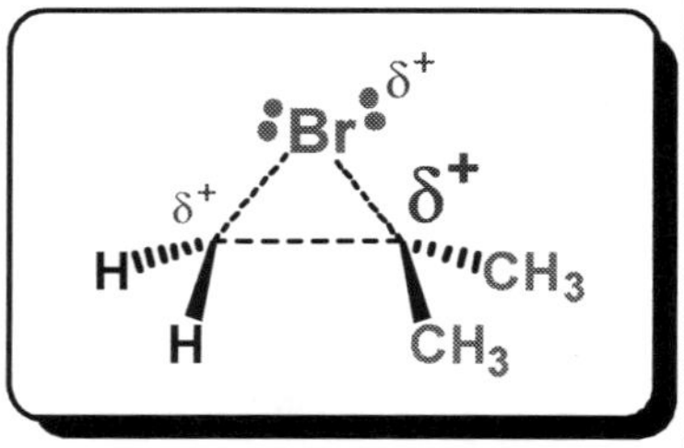


*Resonance hybrid*

(B)

chloride will attack the more positive carbon.

**Notes:**

# Halohydrin Preparation

The halonium ion is also a useful intermediate en route to other products, such as **halohydrins**. In this case, water or alcohol serves as the nucleophile to attack the halonium formed by reaction of $X_2$ with an alkene:

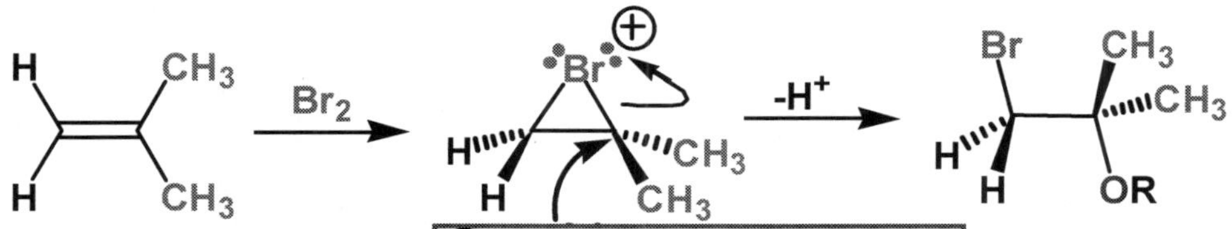


(A)

Notes:

(B) Addition is:

water/alcohol attacks:

(C)

**Notes:**

**Example.** Give correct product(s). Show stereochemistry.

(A)
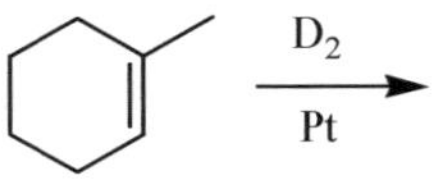


(B)
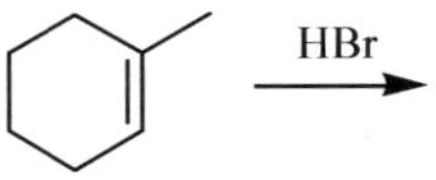


(C)
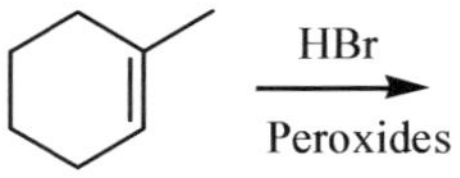


(D)
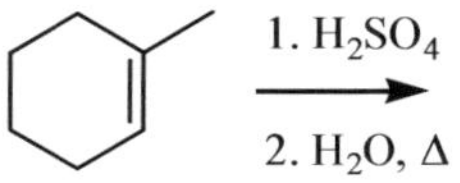


(E)
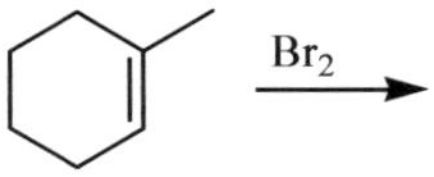


(F)
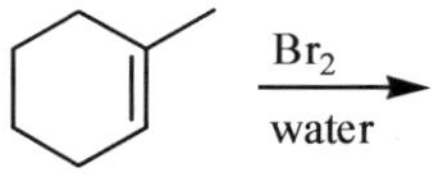


(G)
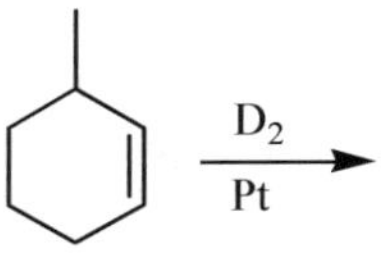


(H)
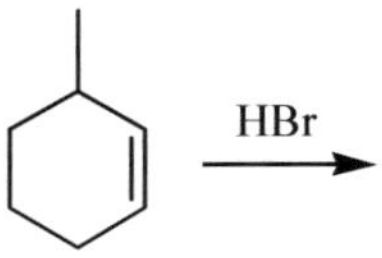


(I)
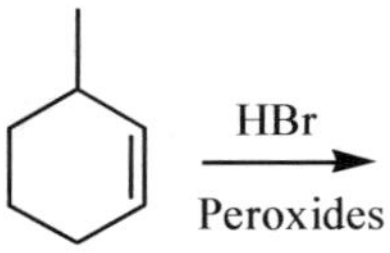


(J)
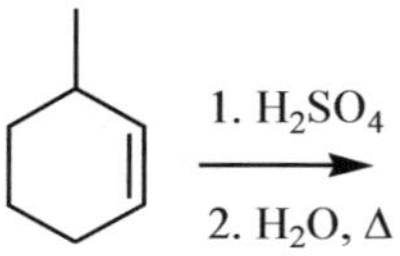


(K)
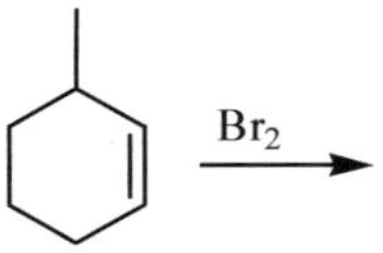


(L)
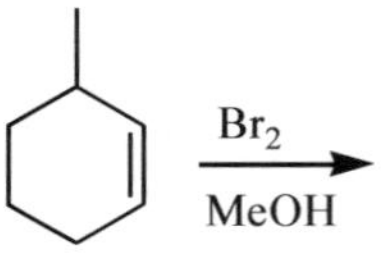


**Notes:**

# Preparing Alcohols: Oxymercuration/Reduction

A mechanism similar to halohydrin preparation can be used to prepare alcohols:

reaction:

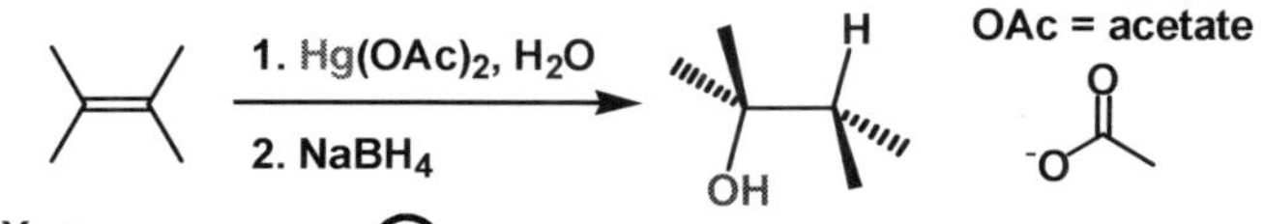


mechanism:

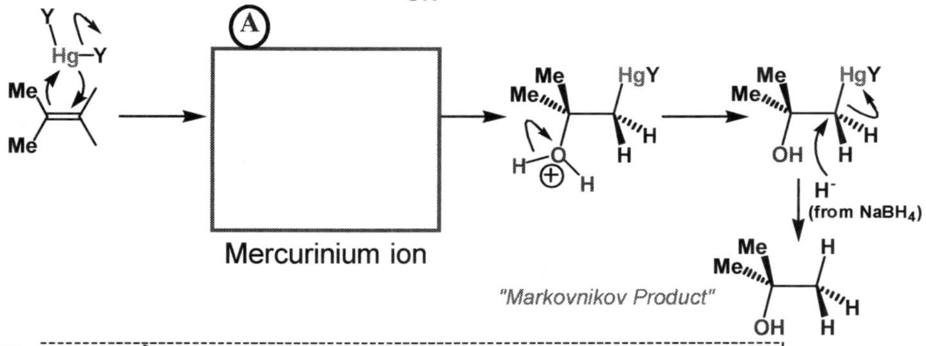


compare to:

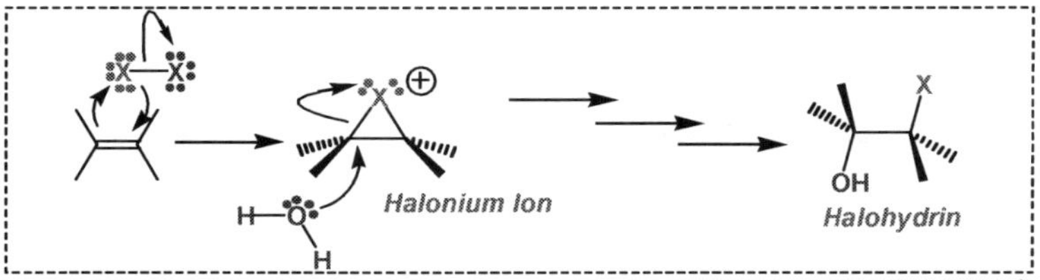


B

Why bother if we can use simple hydration???

## Notes:

# Oxymercuration/Reduction: Regioselectivity

Just as for halonium ion intermediates, we can write various resonance forms for the mercurinium ion. There will be more positive charge on the more substituted carbon, because this is the more stable carbocation:

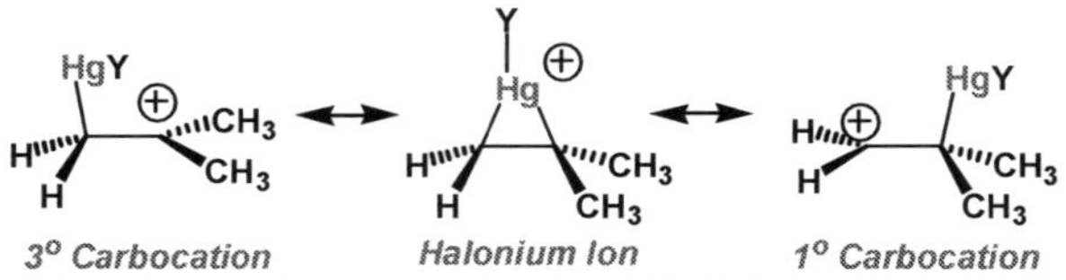


Ⓐ Hybrid:

As a result:

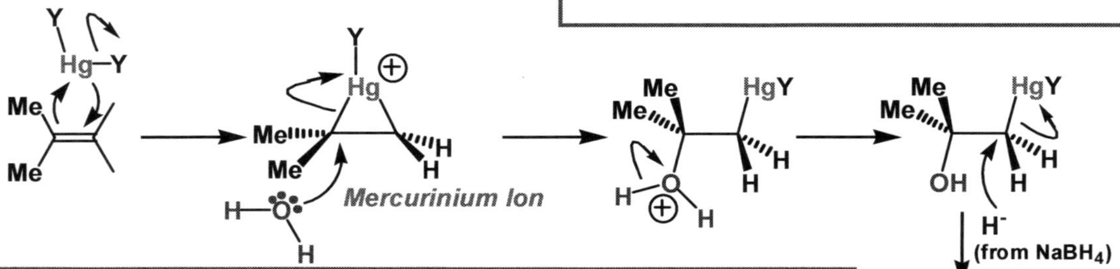


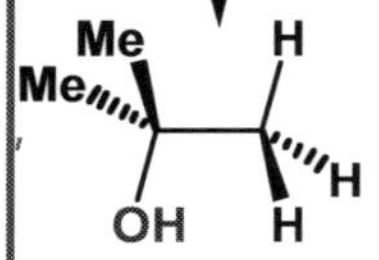


Ⓑ The product:

Ⓒ Addition: Not selective for syn or anti!!!

**Notes:**

# Hydroboration/Oxidation

An alternative protocol can be used to prepare 'anti-Markovnikov' alcohols:

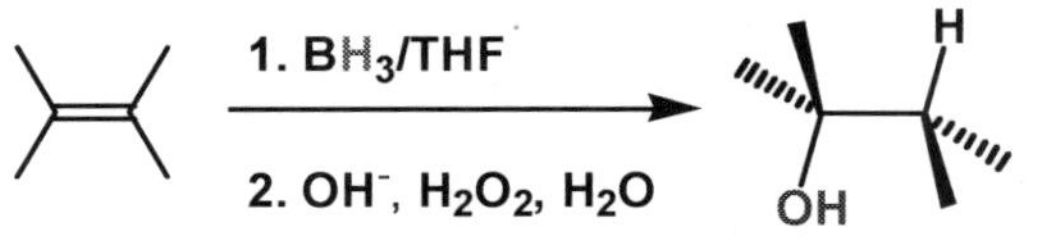


( you may see "$B_2H_6$" or "$R_2BH$" in place of "$BH_3$/THF"; same mechanism!)

(A) The borane:

The Alkene:

(B)

**Notes:**

# Hydroboration/Oxidation Mechanism

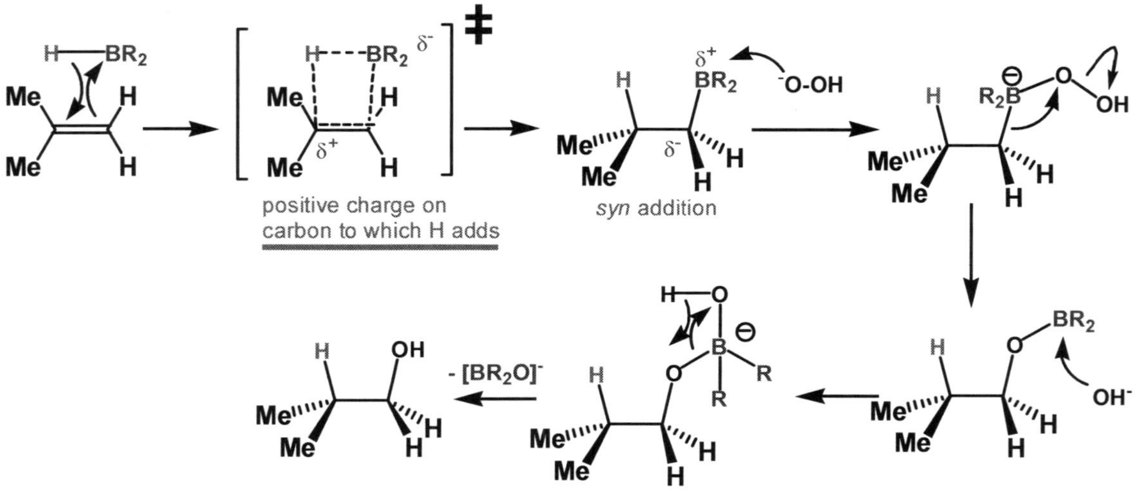


A Addition is:

**Notes:**

(A) Transition state tells what the major product is:

(B)

(C) Why bother? We have hydration and oxymercuration to make alcohols already!:

**Notes:**

# Mechanism Dictates Regioselectivity

Rearrangement in hydration with acid (carbocation) but not with mercurinium or borane intermediates.

(A) 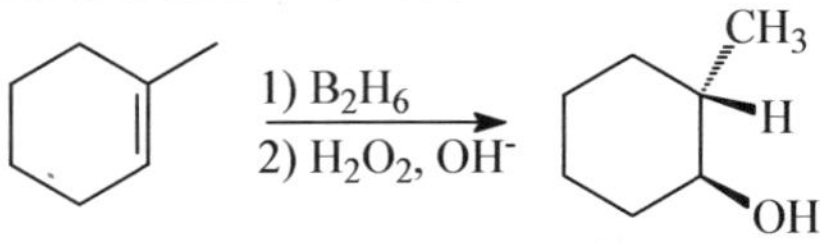


Anti-Markovnikov Product; Syn addition

(B) 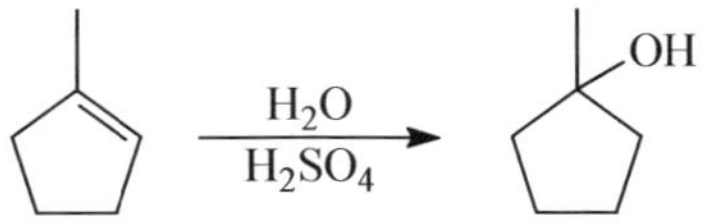


Markovnikov product

(C) 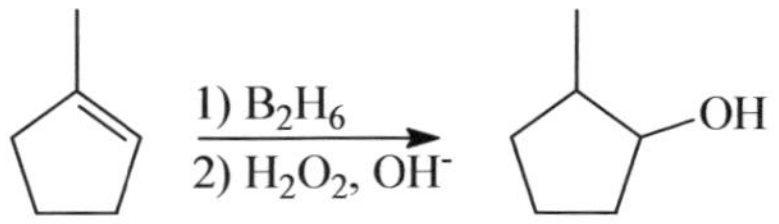


Anti-Markovnikov product; syn addition

(D) 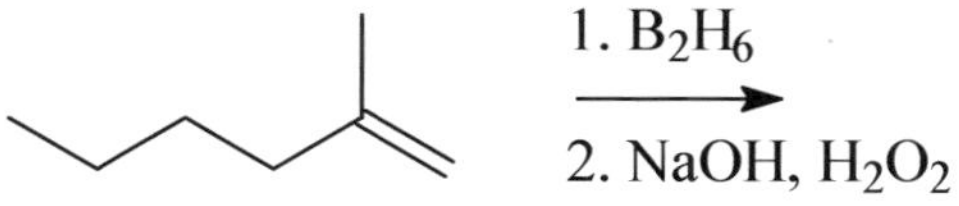


(E) 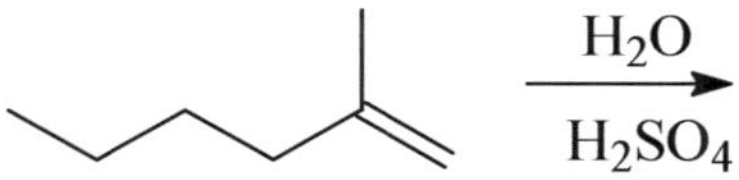


**Notes:**

# Cyclopropanation

We have seen how cationic species can add to an alkene to form three-membered rings. Highly reactive neutral species, such as **carbenes** can react with alkenes to make neutral, cyclic products. One example of such a transformation is the reaction with a carbene or carbenoid to form a cyclopropane derivative:

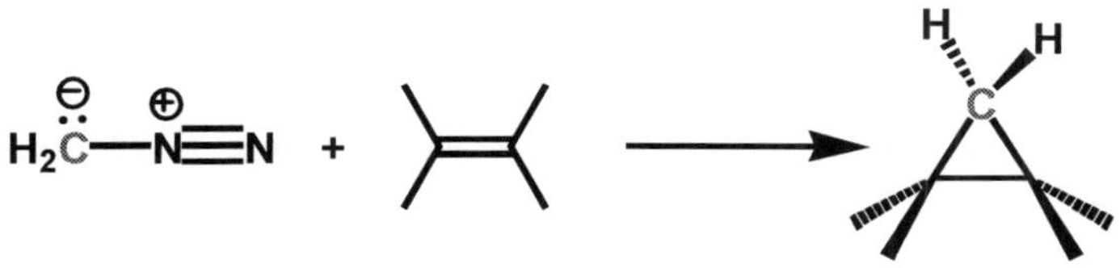


The mechanism:

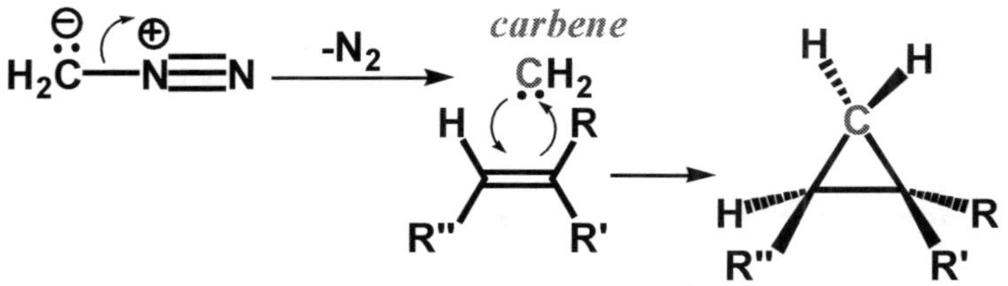


Ⓐ Addition is:

**Notes:**

# Cyclopropanation

A strong base (such as $KO^tBu$) can remove HCl from chloroform ($CHCl_3$) to generate dichlorocarbene:

(A)

This carbene can then add to an alkene to make a dichlorocyclopropane:

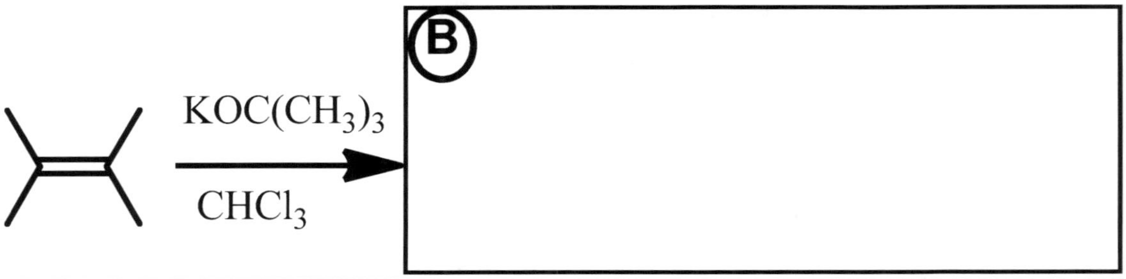


**Notes:**

When Zn(Cu) and $CH_2I_2$ react, a **carbenoid** results:

(A)

This carbene can then add to an alkene to make a dichlorocyclopropane:

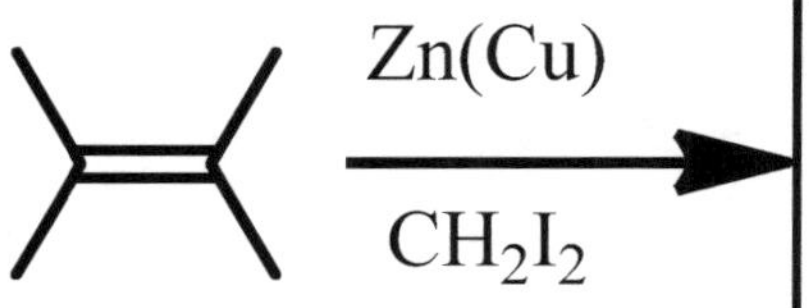


(B)

This route of cyclopropanation is called the

(C)

**Notes:**

<u>Carbeneoid:</u>

# Epoxidation

A three-membered ring with one oxygen atom in the ring is called an **epoxide**. Epoxides can be prepared from alkenes:

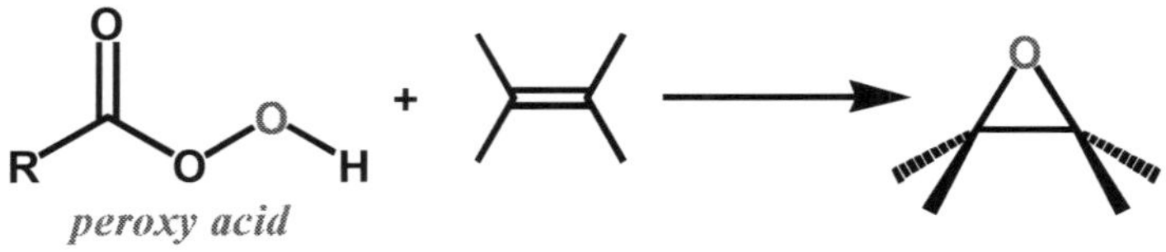

*peroxy acid*

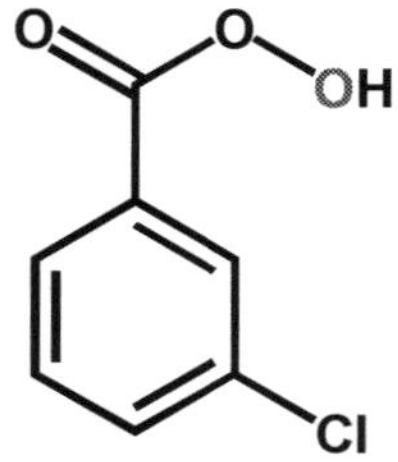

*m*-chloroperoxybenzoic acid
(mCPBA)

The mechanism:

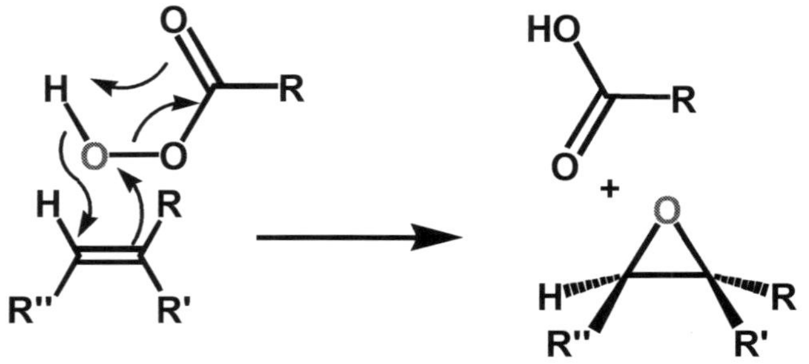

Ⓐ Addition is:

**Notes:**

# Naming Epoxides

To name an epoxide, add the 'epoxy' prefix to the parent chain, and indicate the two carbon numbers to which the epoxide oxygen is bound:

Epoxyethane

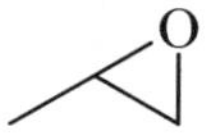


1,2-epoxypropane

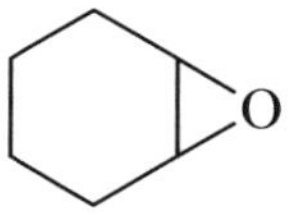


1,2-epoxycyclohexane

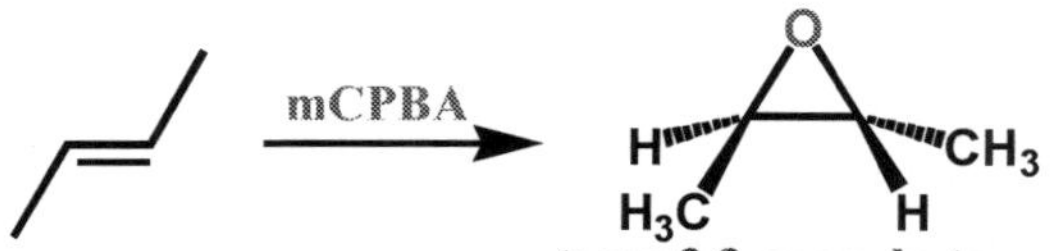


*trans*-2,3-epoxybutane

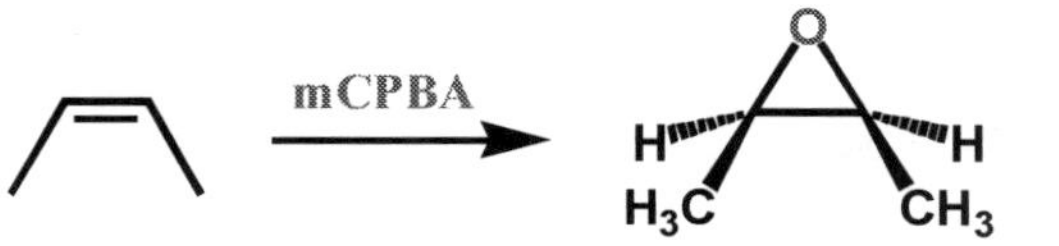


*cis*-2,3-epoxybutane

A

**Notes:**

**Example.** Give correct product(s). Indicate stereochemistry.

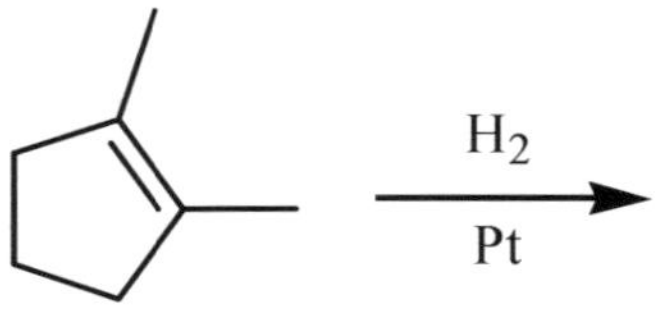


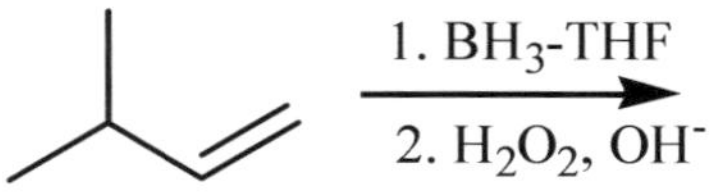


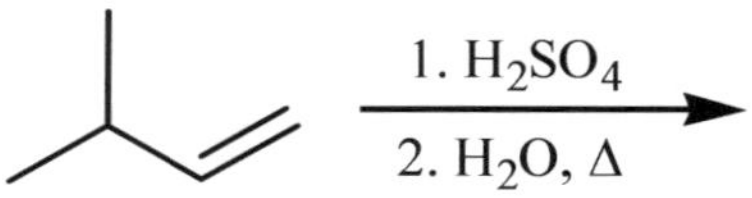


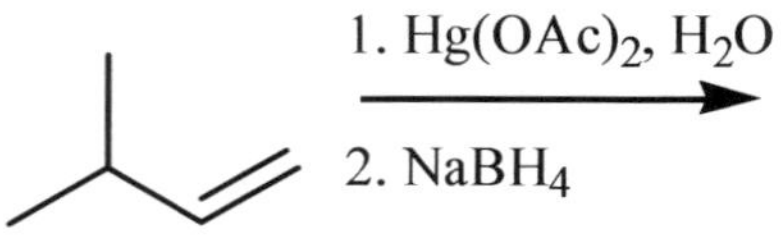


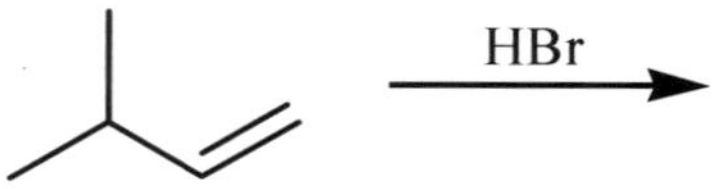


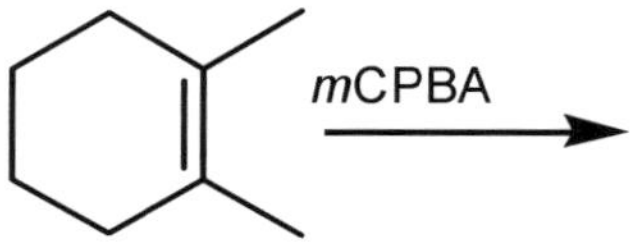


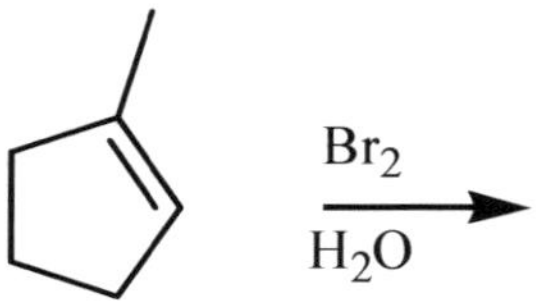


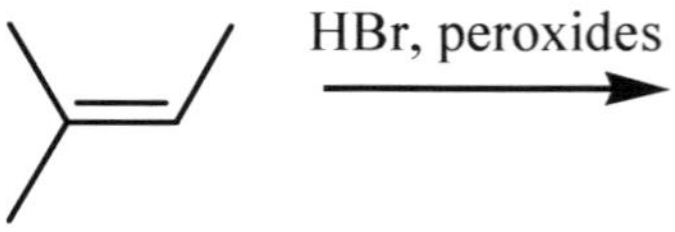


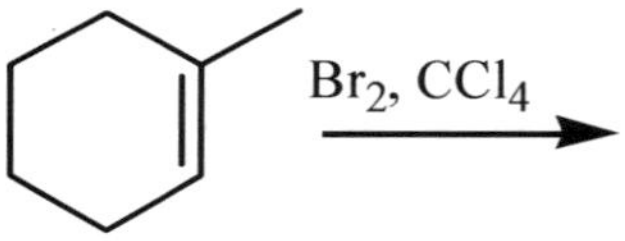


**Notes:**

**Example.** Show reagents/conditions that give the target products.

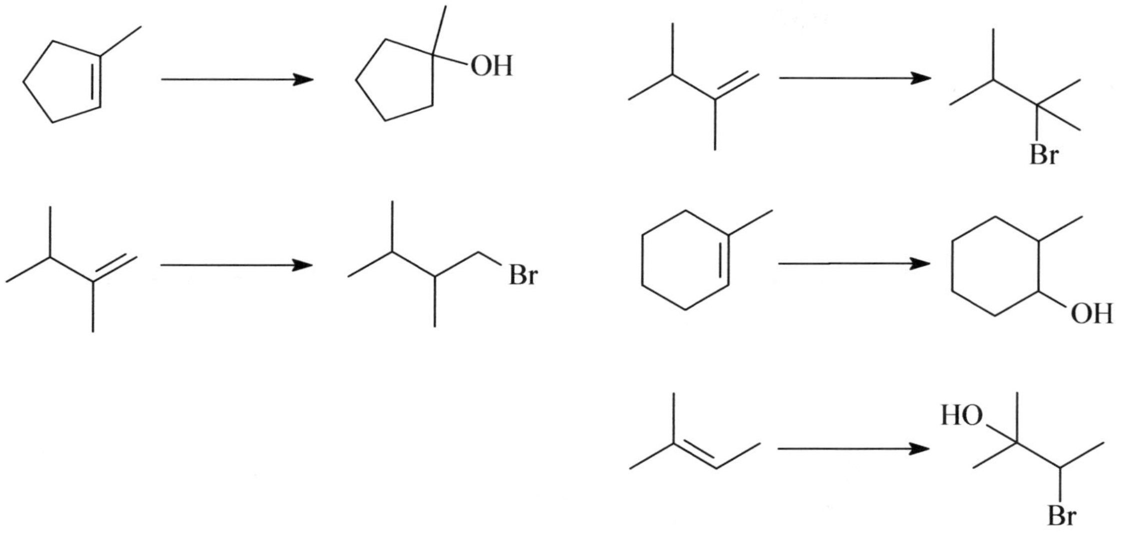


**Notes:**

# Oxidation to Vicinal Diols

Organic Chemist Definition of oxidation and reduction::

(A)

Reaction of an alkene with osmium tetroxide ($OsO_4$) followed by a reductive workup leads to net *syn*-addition of two OH groups to the two carbons that were in the C=C bond:

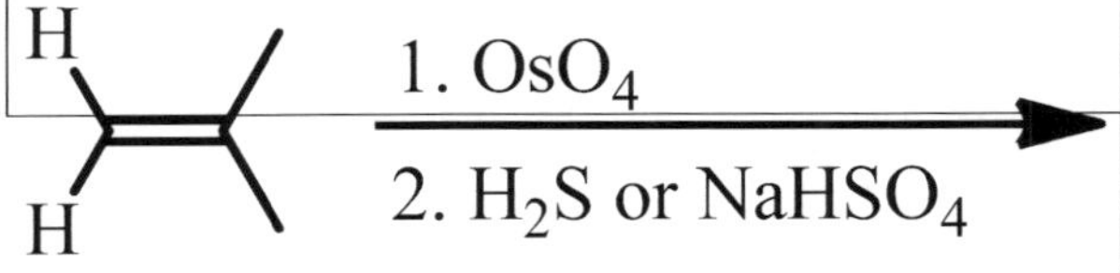


(B)

An example where *syn*-addition would be evident:

(C)

**Notes:**

# Ozonolysis

Ozonolysis differs from the other reactions of alkenes in that instead of *adding* atoms to the pi bond, the net result of this reaction is that the double bond is *broken*.

The first step of ozonolysis is to react the alkene with ozone ($O_3$):

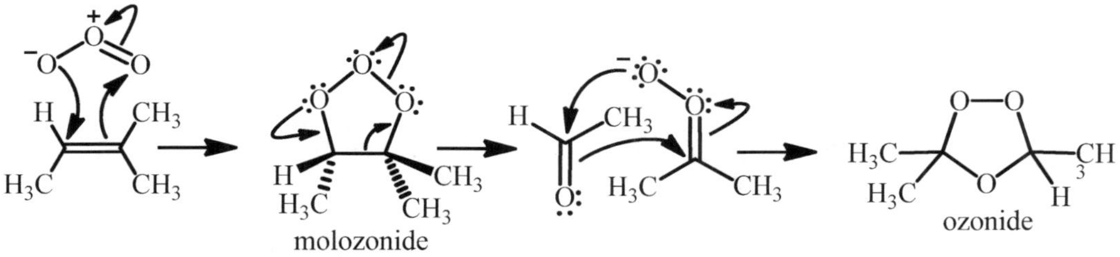


There are two different workup conditions one can use after reaction with ozone:

**(A)** Reducing conditions workup ($Zn/H_2O$ or $R_2S$):

**(B)** Oxidizing conditions ($H_2O_2$)

**Notes:**

# Ozonolysis

The net reactions that you see will look like this:

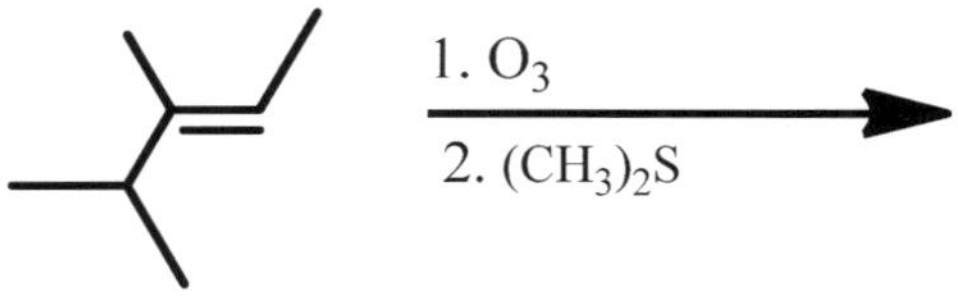


A

B

If there is more than one double bond, you can break all of the C=C bonds to make carbonyls:

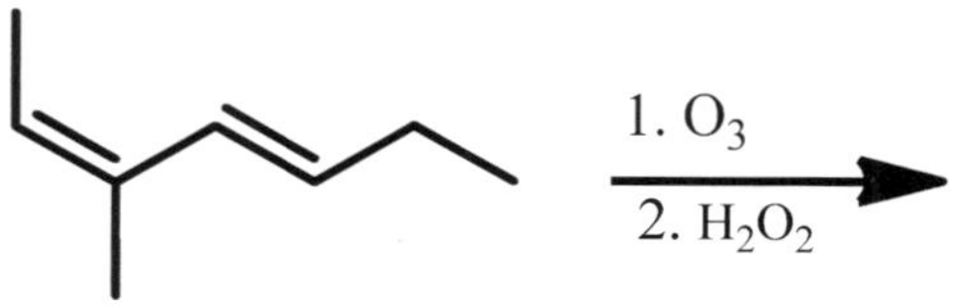


C

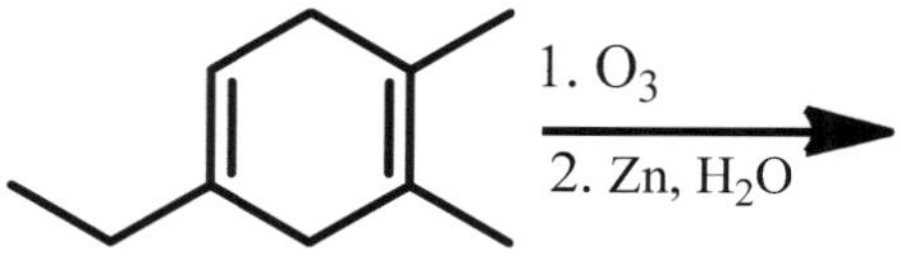


D

# Polymerization

Alkenes can be used as starting materials to make **polymers**.

**POLYMER:**

**MONOMER:**

**REPEAT UNIT:**

**EXAMPLE:**

**Notes:**

# Polymerization

There are three ways that one can polymerize an alkene. The key initial species generated can be a cation, a radical, or an anion.

**Cationic Polymerization of an alkene:**

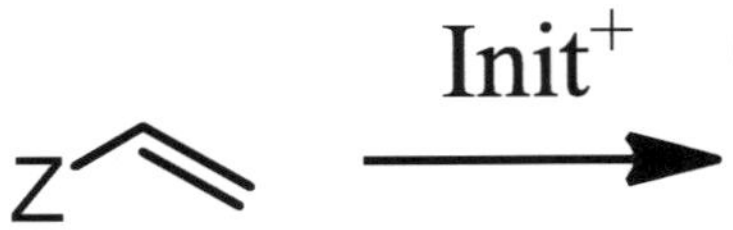


**A**

**B**

Lewis acids can be good initiators for a cationic polymerization:

**C**

**Notes:**

# Polymerization

**Radical Polymerization of an alkene:**

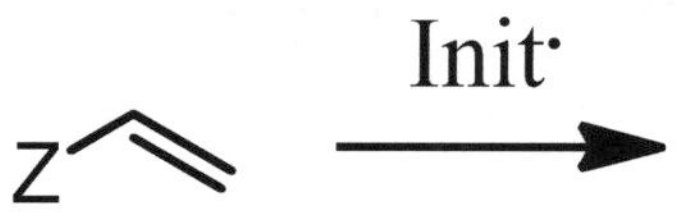


A

B

Unlike the other radical reactions of alkenes that we have seen, there is ONLY an initiator added to the alkene, no other reagents. A common radical initiator is Benzoyl Peroxide:

C

**Notes:**

# Polymerization

There are three ways that one can polymerize an alkene. The key initial species generated can be an anion, a cation, or a radical.

**Anionic Polymerization of an alkene:**

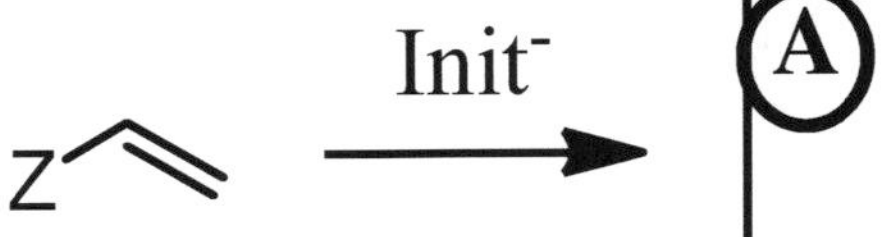


B

You may notice that in the other reactions of alkenes in this unit, a cation or radical was the key intermediate. This is because it is more difficult to generate a carbanion. In order to do a successful anionic polymerization, **the group Z must**:

C

**Notes:**

# Polymerization

Here are some examples of important polymers made by the polymerization of alkenes:

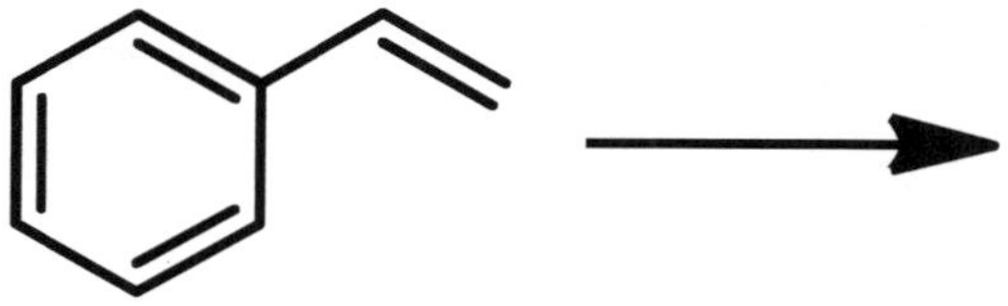

(A) polystyrene

(B)

polyethylene

(C)

poly(methyl methacrylate)

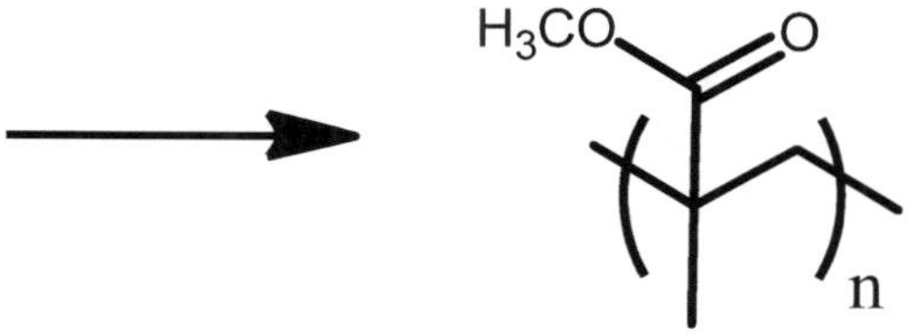


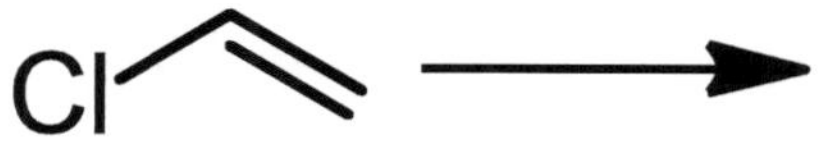


(D) Poly(vinyl chloride) (PVC)

# Lecture Set 6: Stereochemistry

**Suggested Reading:**

**Suggested Problems:**

**Stereoisomers**:

(A)

**Conformational stereoisomers**:

(B)

- staggered and eclipsed conformers
- *gauche* and *anti* conformers
- Chair or boat chair conformers of cyclohexane

**Configurational stereoisomers**:

(C)

*cis* and *trans* isomers    *E* and *Z* isomers

**NOW:**

**enantiomers**:

(D)

**diastereomers**:

(E)

# Notes:

# Chirality

Molecules with "handedness" possess (A)

and are (B)

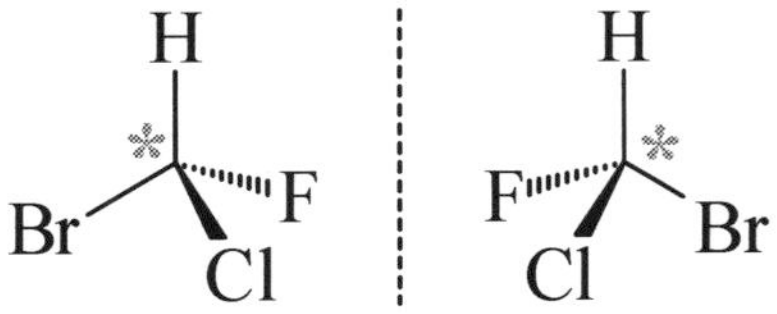


These are nonsuperimposable mirror images; using definitions on the previous slide, this is a pair of **enantiomers** ("left- and right-handed" isomers). The molecule is chiral and possesses a **chiral** (**stereogenic**) **center** (also called a **stereocenter**). These centers can be labeled with asterisks for clarity.

(C)

**Notes:**

(B)

Mirror images are **superimposable** if you rotate it and see it is the same molecule.

A molecule WITHOUT a stereocenter is

(B)

If the molecule has a **plane of symmetry** it is

(C)

## Notes:

## Stereocenter?

Example. Mark the stereocenters with asterisks.

(A)

(B)

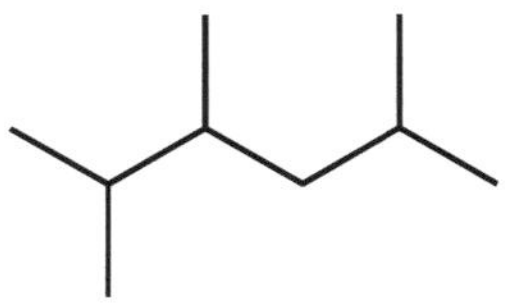

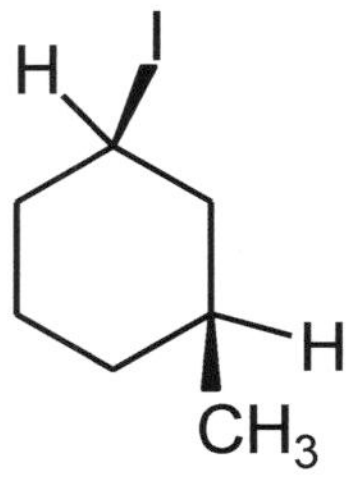


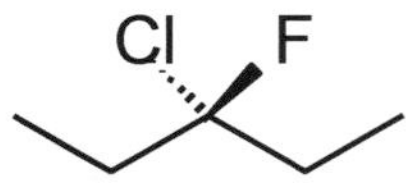


**Notes:**

# Chiral?

Example. Mark stereogenic carbons with asterisks. If the molecule has a chiral center, draw both enantiomers using wedge/dash notation.

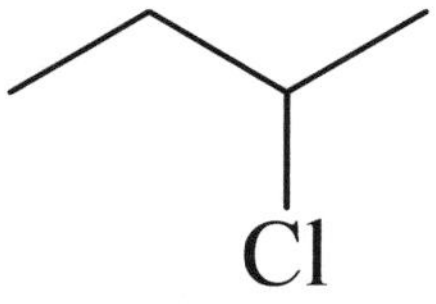


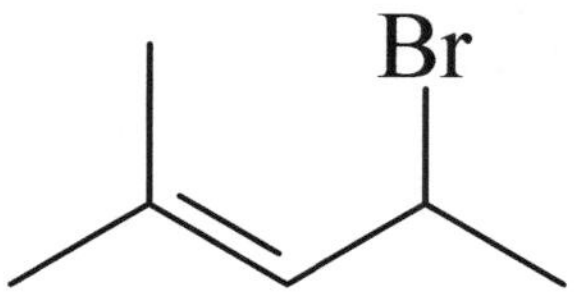


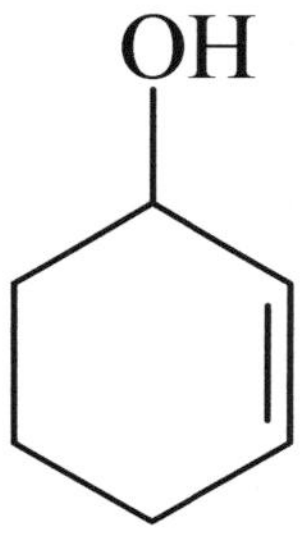


**Notes:**

# Properties of Enantiomers

Many physical properties of enantiomers are **IDENTICAL**

(A)

If all these physical properties are the same, what is the importance of studying the chirality of a molecule?

(B)

1. Amino acids

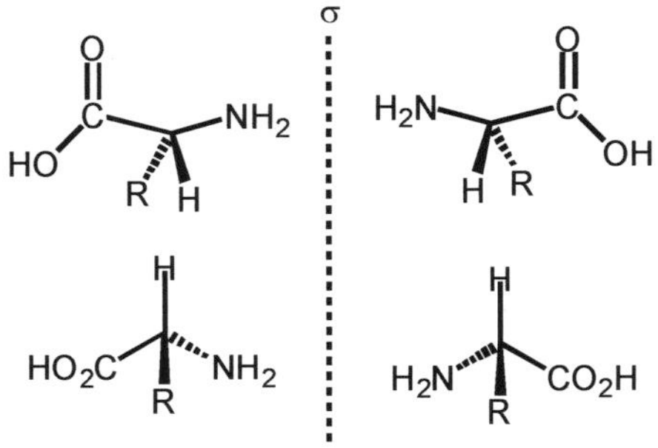


**Notes:**

# Properties of Enantiomers

2. Enzymes/cellular signaling

A

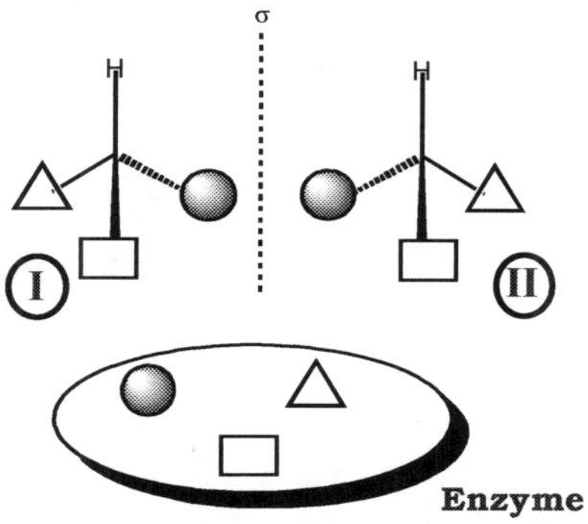


Enzyme or Cellular Receptor

3. Drugs

B

**Notes:**

The handedness of a molecule is vitally important. But many physical properties we use to distinguish different compounds will be the same for both enantiomers. How can we tell them apart???

(A)

(B)

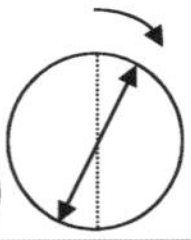

(C)

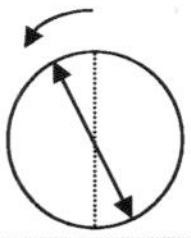

(D)

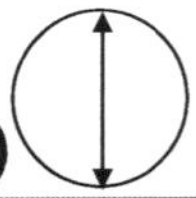

**Notes:**

# Optical Activity Terms

**Optically inactive**: No rotation of polarized light; indicates:

(A)

**Optically pure**: (C)

**Optical purity:** (B)

(a sample with 80% enantiomer '**X**' and 20% enantiomer '**Y**' has 60% e.e. or 60% optical purity)

**Notes:**

# Optical Activity:

Enantiomers of different compounds rotate light by different amounts, so an optically pure compound has a characteristic specific rotation [α]. Two enantiomers of the same compound have **specific rotations** with the same magnitude but in opposite directions. The equation used to calculate the specific rotation is:

$$[\alpha]_D^T = \frac{\alpha}{cl}$$

temperature

observed rotation (°)

path length (dm)

Type of light used
(sodium D line)

concentration (g/mL sol'n)

(+)-2-butanol $[\alpha]_D^{27}$ = + 13.5°
(–)-2-butanol – 13.5°

Ⓐ **nomenclature***

**Notes:**

# Absolute Configuration

We now know that the handedness of a molecule is vitally important and that there are physical property differences we can use to distinguish enantiomers from one another. We obviously need a nomenclature to designate the **absolute configuration** of the molecule.

(A)

(B)

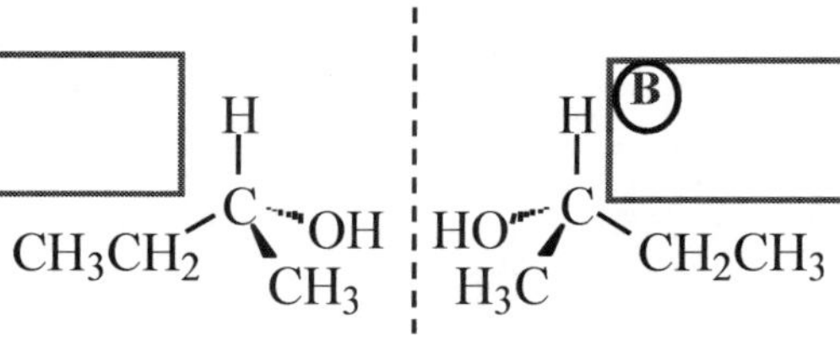


The absolute configuration is designated as *R* or *S*.

Whether the configuration of a chiral center is *R* or *S* depends only on

(C)

and is **NOT** determined by

(D)

R: *rectus* (Latin "right")    S: *sinister* (Latin "left")

**Notes:**

# Assigning Absolute Configuration: Newman

1. (A)

   (see the "**Easy EZ**" section); priority rules are often referred to as the Cahn-Ingold-Prelog rules with reference to chirality.

2. (B)

3. Determine the direction of procession of the other three groups, from **highest to lowest**:

   clockwise = (C)

   counterclockwise = (D)

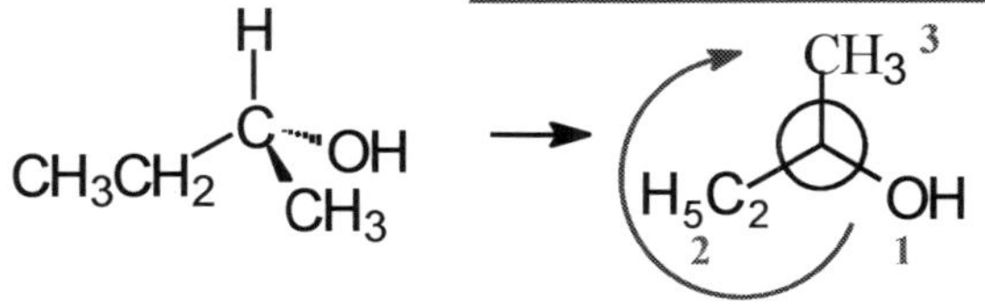


(E)

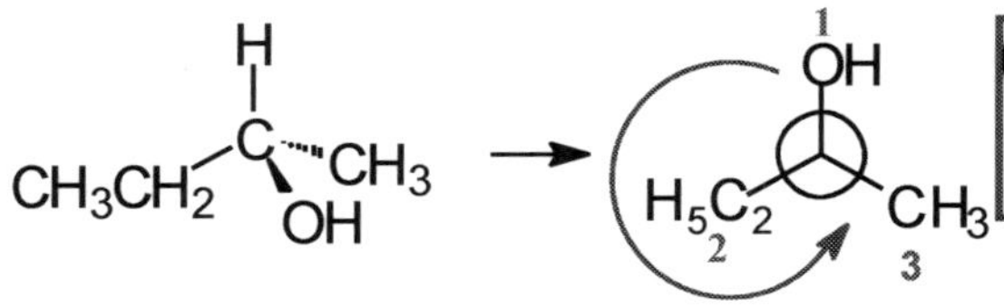


(F)

**Notes:**

# Absolute Configuration Question

Example. Name these compounds.

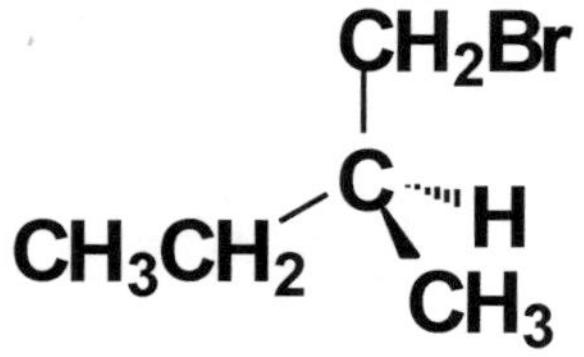


(A)

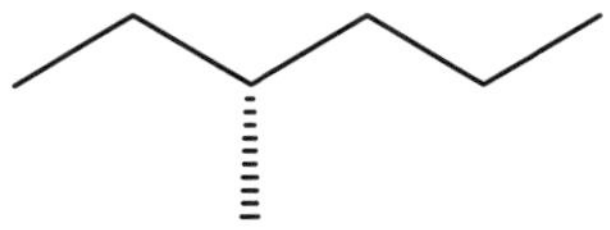

(B)

**Notes:**

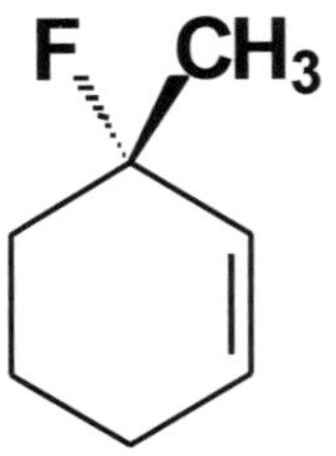


(C)

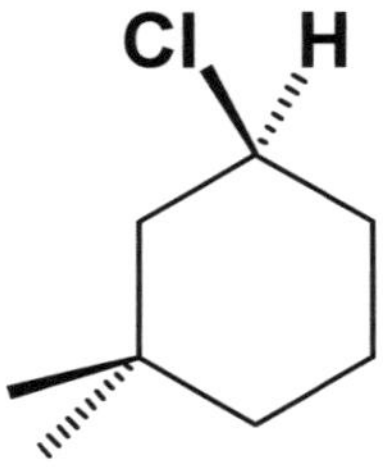


(D)

**Notes:**

# Fischer Projections

In addition to standard dash-wedge notation, there is another very useful manner in which to represent the 3D shape of a molecule called a **Fischer projection**:

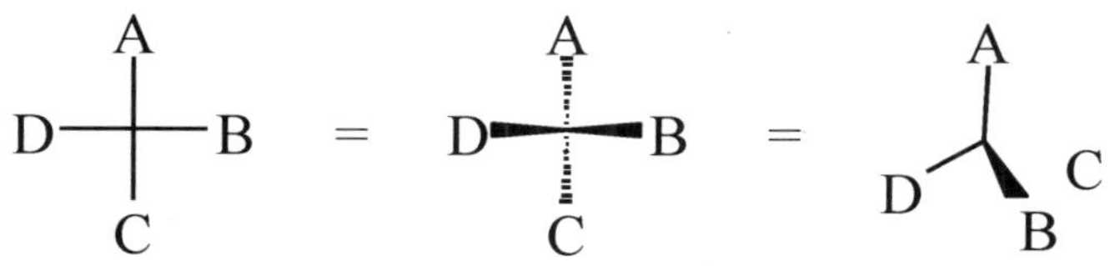


(A)

(B)

**Notes:**

Determining *R* and *S* in Fischer projections:

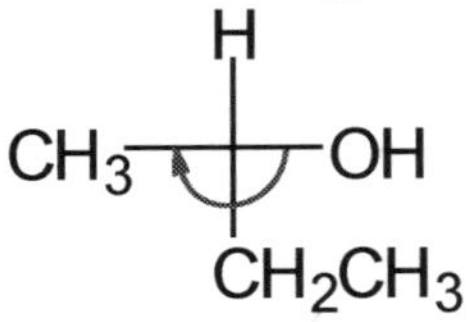


(A)

Convert to wedge and dash structure:

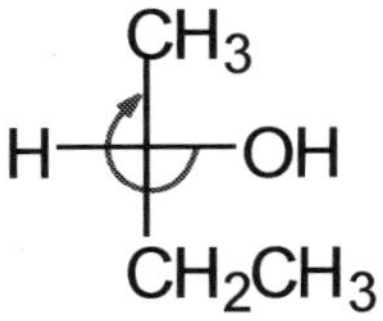


(B)

Convert to wedge and dash structure:

**Notes:**

# Reactions and Chirality: Radical Halogenation

Some of the reactions we have discussed produce chiral centers.

(A)

**OBSERVE:**

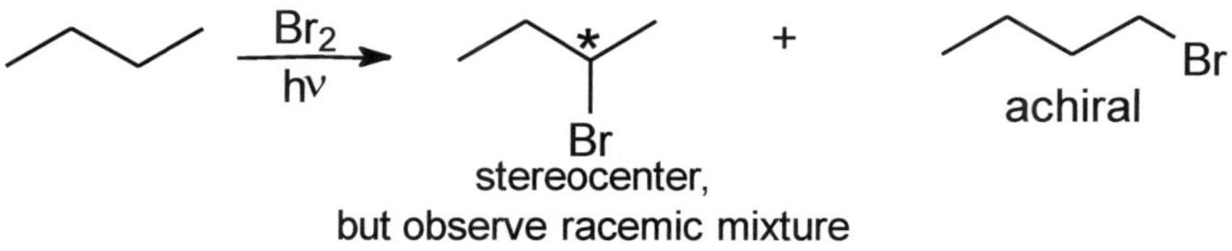


**EXPLAIN:**

(B)

**Notes:**

# Reactions and Chirality: Hydration

**OBSERVE:**

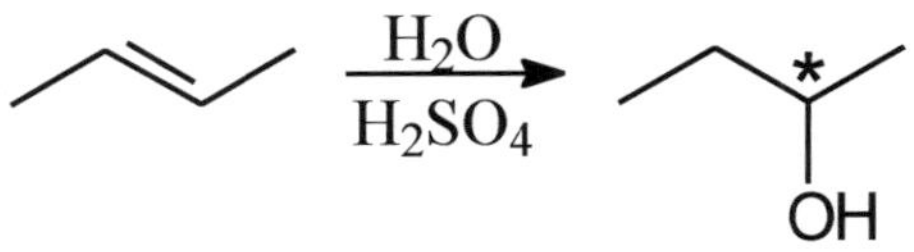


Has a stereocenter but a racemic mixture is observed

**EXPLAIN:**

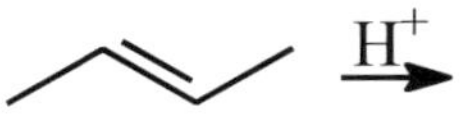


A

**Notes:**

# Reactions and Chirality Problem

Example. Show all products including their stereochemistry.

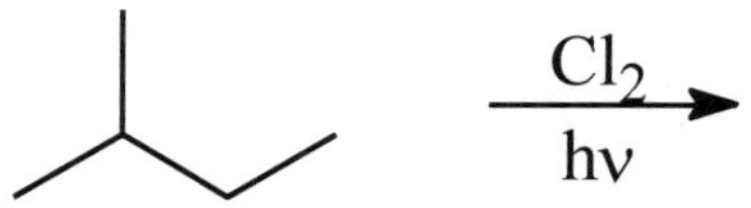


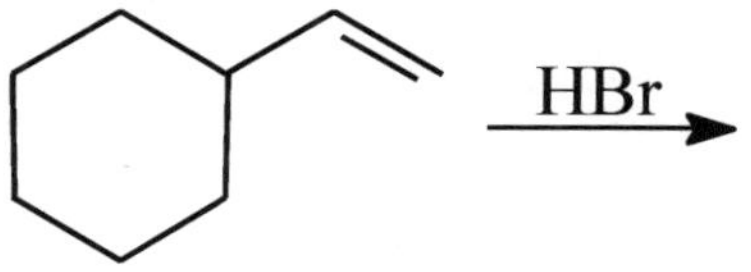


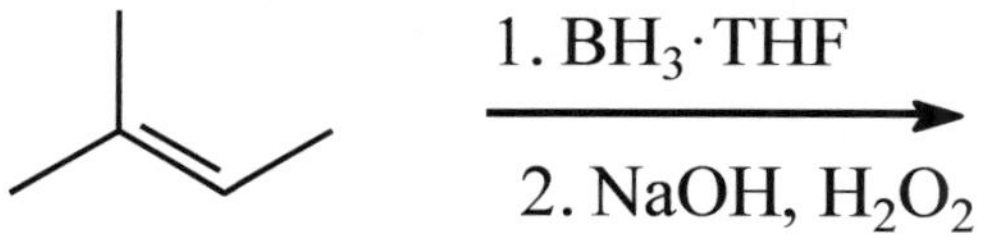


**Notes:**

# Multiple Stereocenters: Diastereomers

diastereomers:

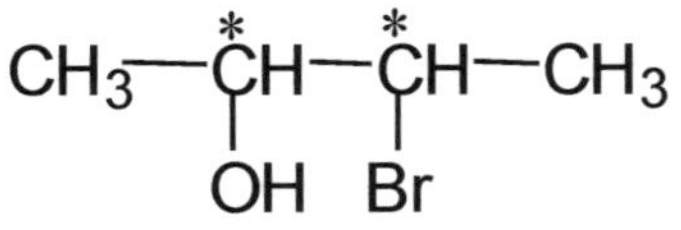

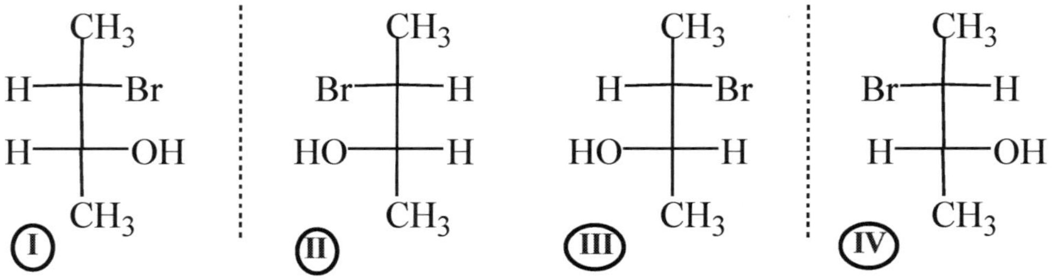


**Notes:**

# Meso Compounds Problem

Example. Draw all stereoisomers of 3,4-dibromohexane using Fischer projections. Assign absolute configuration to each stereocenter. Denote stereochemical relationships between each pair.

**Notes:**

# Stereoisomers of Substituted Cyclohexane

Ⓐ

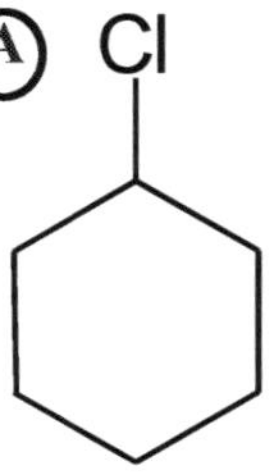


Ⓑ

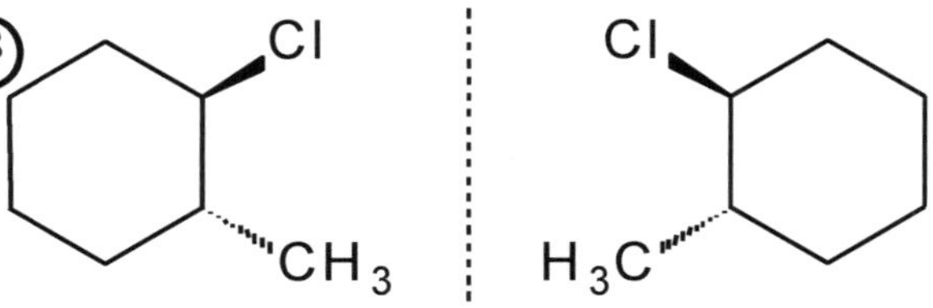


Ⓒ

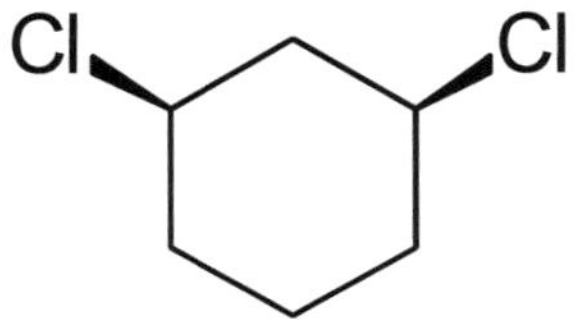


Ⓓ

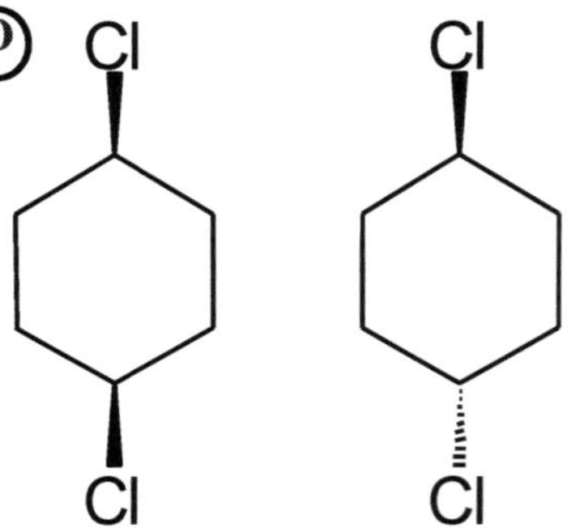


**Notes:**

# Stereoisomers of Substituted Cyclohexane

Example. Assign absolute configuration at the stereocenters. Label each compound as chiral or achiral. If a plane of symmetry is present, label it.

(A)

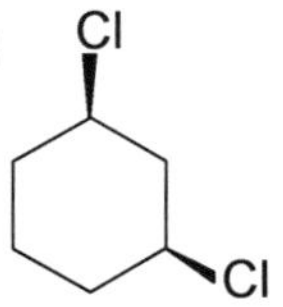


(B)

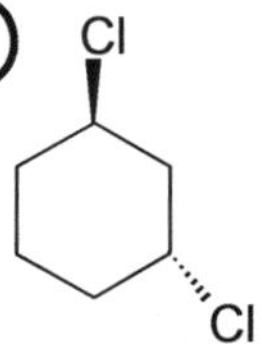


(C)

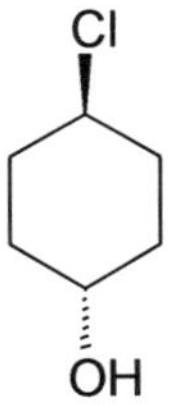


(D)

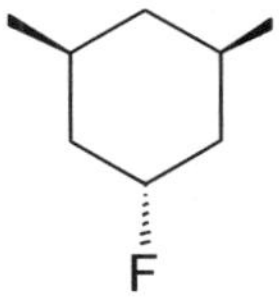


**Notes:**

# Stereoisomer Problem

Example. Designate each pair as identical, enantiomers, or diastereomers.

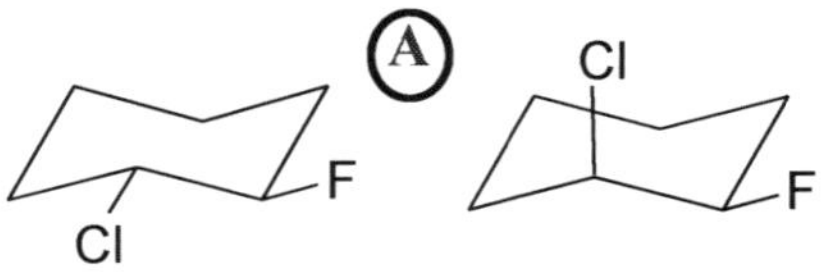


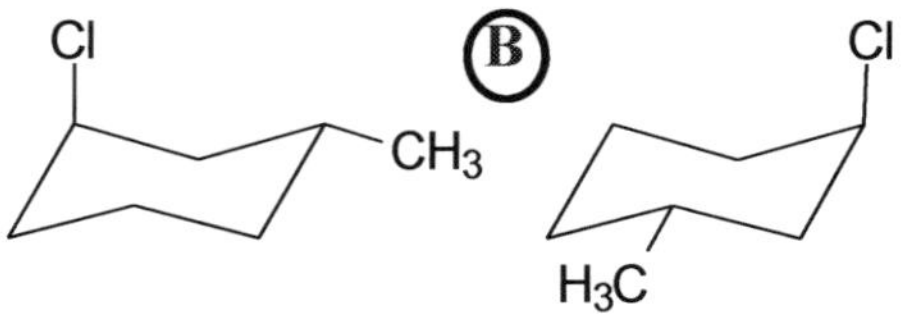


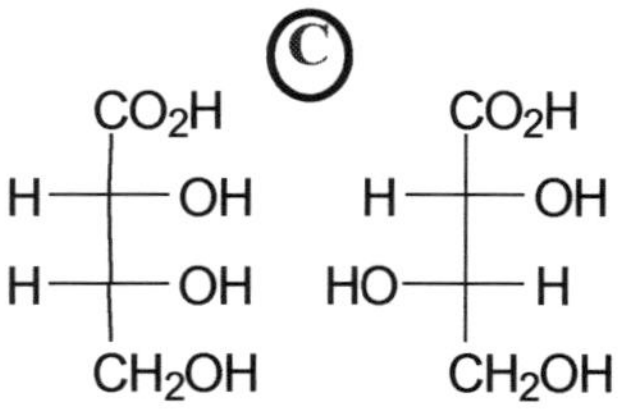


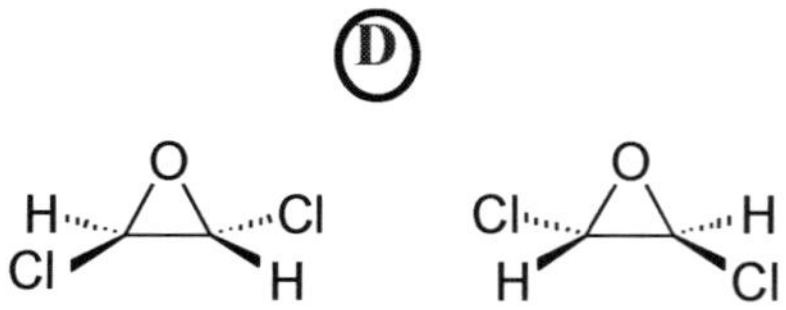


**Notes:**

# Reactions giving Diastereomers

Ⓐ **stereospecific** reactions:

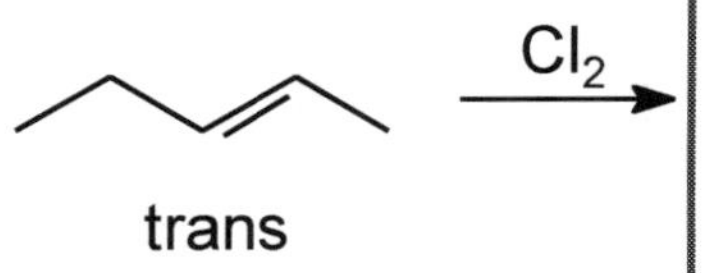


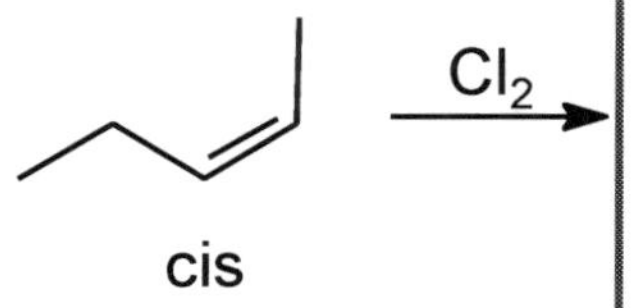


**Notes:**

# Reactions giving Diastereomers

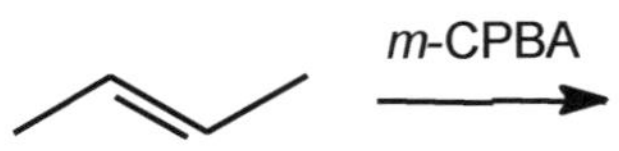


A

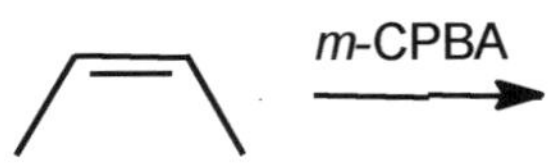


B

**Notes:**

# Reactions giving Diastereomers

We know that you cannot get optically active products from optically inactive starting materials. However, if the starting material is optically active, it is possible that one of multiple possible stereoisomeric products will be favored:

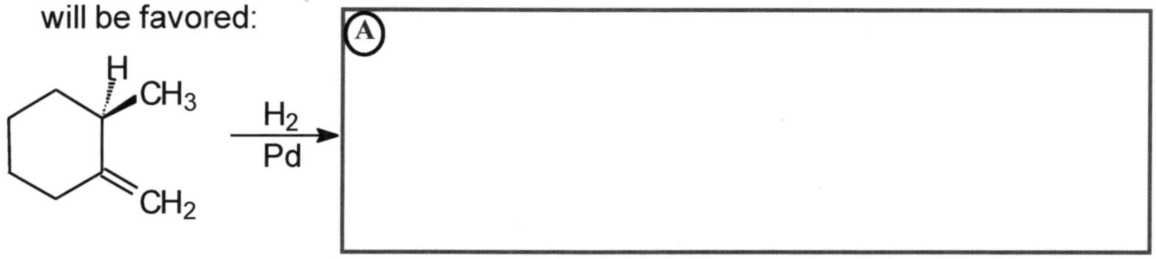


70% 30%

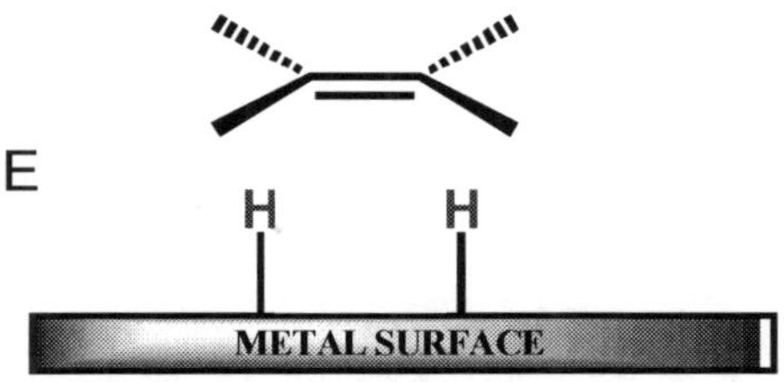


Why do we observe this ratio?
Recall that the reaction happens AT THE SURFACE:

**Notes:**

# Lecture Set 7:
# Alkynes

**Suggested Reading:**

**Suggested Problems:**

# Naming Alkynes

acetylene — terminal alkyne — internal alkyne

(A)

Alkanes: saturated hydrocarbons

Alkenes: unsaturated hydrocarbons (double bond in an alkane)

Alkynes: unsaturated hydrocarbons (triple bond in an alkane)

Nomenclature of alkynes is much like that for alkanes, with modifications:

1. always designate the longest chain having
   (B)
2. Number the parent chain such that
   (C)
3. change the 'ane' ending of the name you would use if it were an alkane into an 'yne' ending to indicate that it has a triple bond in it.

**Notes:**

# Naming Alkynes

propyne

(A)

(B)

1-pentyne

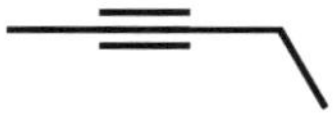
2-pentyne

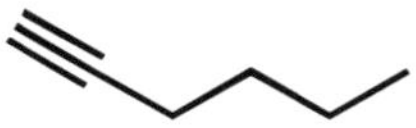
1-hexyne

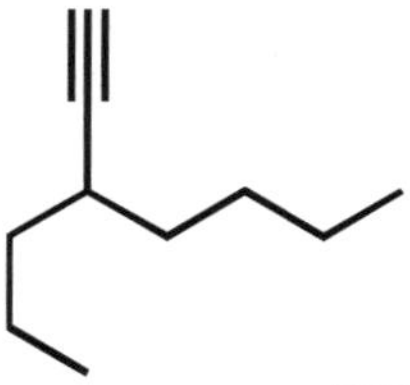
(C)

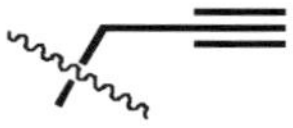
propargyl group

NOTE: Because it is linear there are no E or Z isomers

**Notes:**

4. If there is more than one triple bond, use a "-diyne", "-triyne", etc. at the end of the name in place of the "yne", with multiple numbers indicating where the triple bonds are (just like for alkenes):

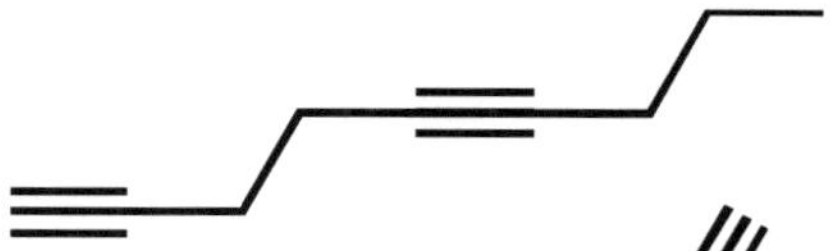

(A)

(B)

5. If alcohol functionalities are present in addition to alkyne:

(C)

if both alkyne and alkene units are present, number the parent so that

(D)

IF AND ONLY IF there is a tie between the two in term of numbering the parent, alkene > alkyne (the **lowest** number is given to the one that comes **first** alphabetically).

**Notes:**

# Naming Alkynes

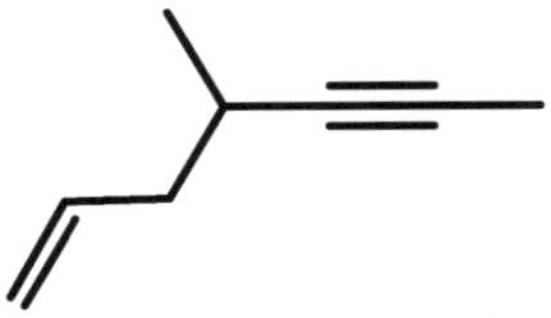

(A)

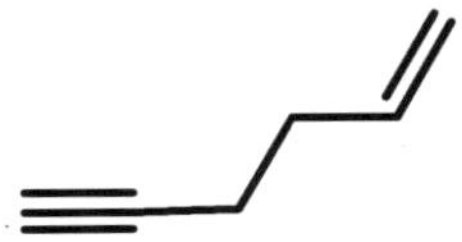

(B)

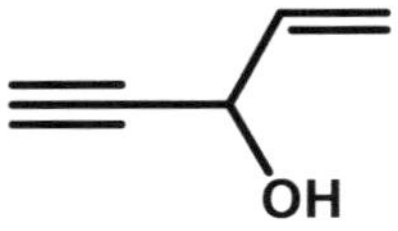


(C)

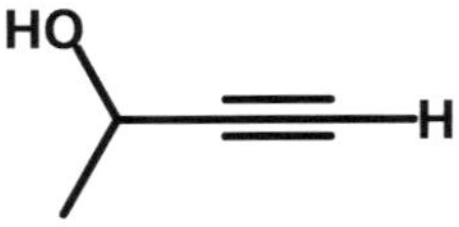


(D)

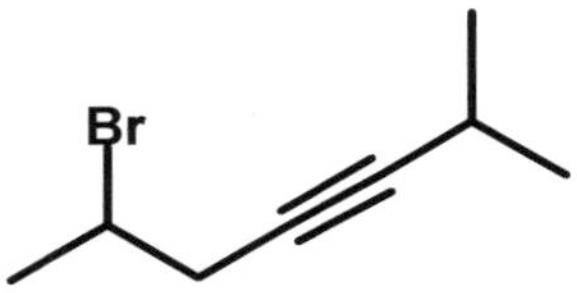


(E)

Note the order of the endings in example C: en-yn-ol

**Notes:**

# Addition to Alkynes

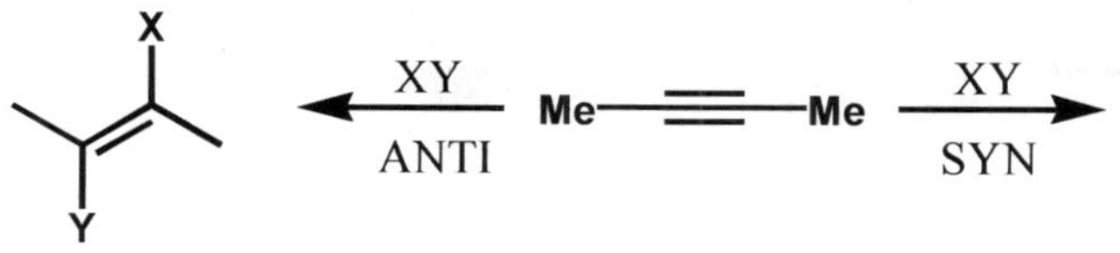


Addition reactions we studied for alkenes can work for alkynes:

(A)

We have TWO pi bonds in an alkyne unit:

(B)

**Notes:**

# Hydrogenation: Alkane or *cis*-Alkene

An alkyne can be hydrogenated using conditions similar to those used to hydrogenate an alkene ($H_2$/ transition metal catalyst):

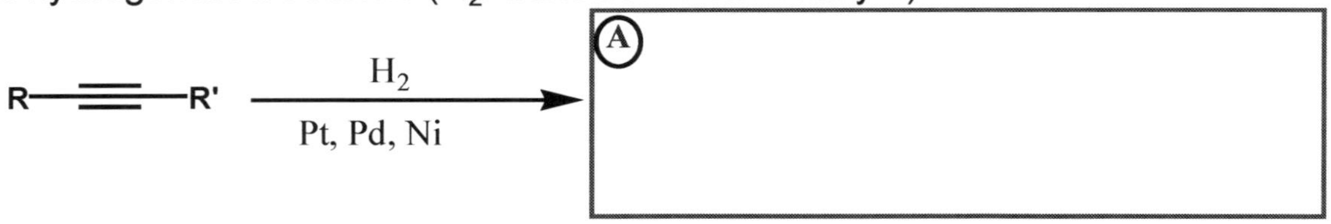


Gentler conditions (a less active catalyst) are required if conversion of an Alkyne to an alkene is desired without further hydrogenation. To do this, use $H_2$ with Lindlar's catalyst ($Pd/CaCO_3$, $Pb(OAc)_2$, quinoline).

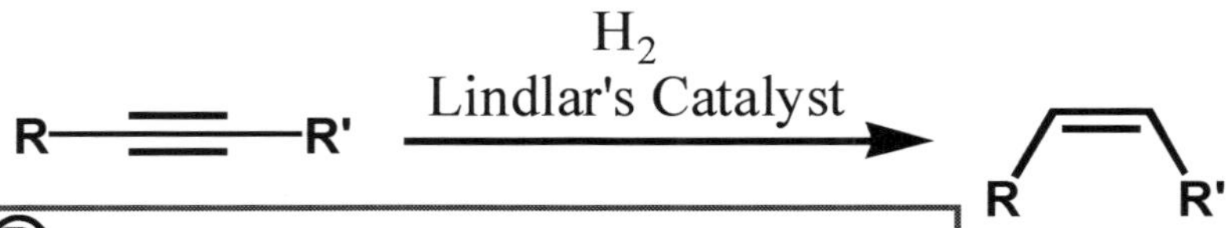


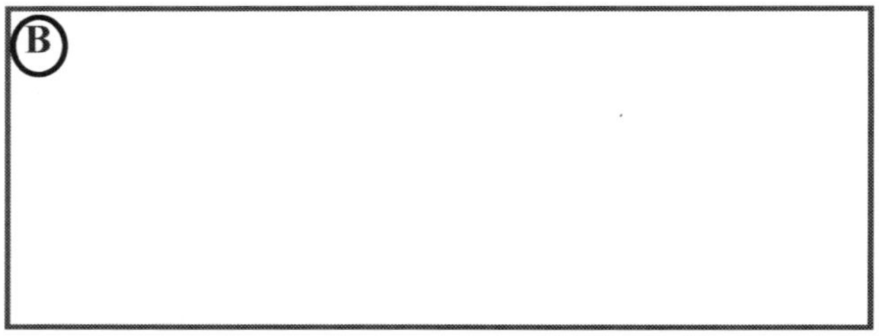


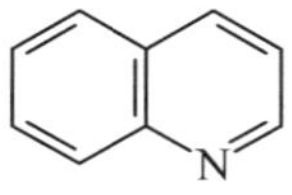


*quinoline*

**Notes:**

# Addition of Hydrogen to Alkynes: *trans*-Alkene

When an alkali metal such as Na or K is dissolved in liquid ammonia, solvated electrons are formed. The resulting solution can be used to reduce alkynes:

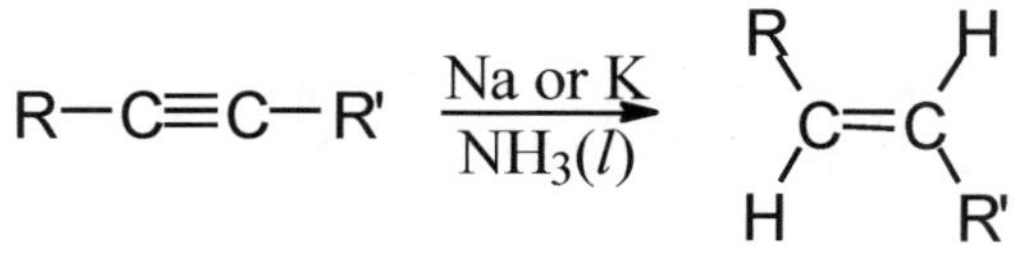


Ⓐ

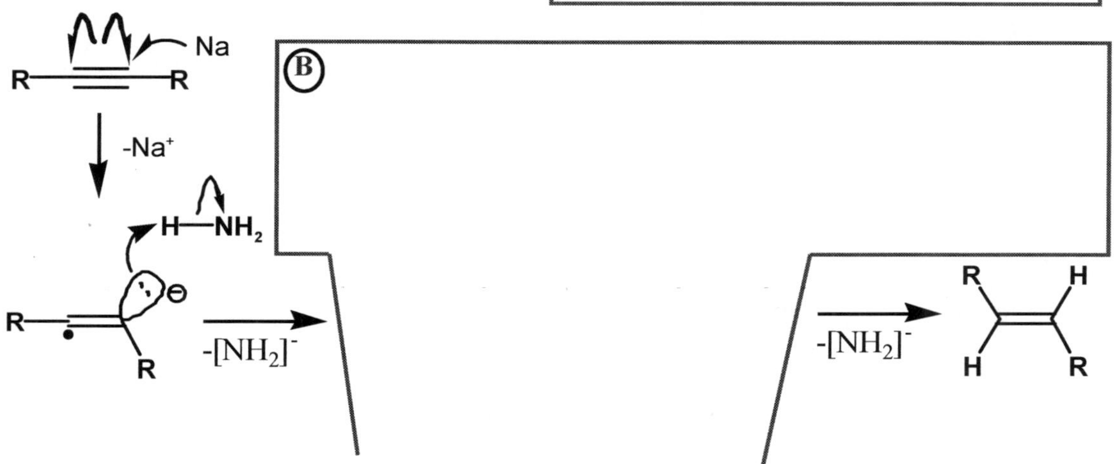


key step

**Notes:**

# Addition to Alkynes: HX

Alkynes can also react with HX, either with or without peroxides, yielding Markovnikov or "anti-Markovnikov" products. You should be able to deduce the mechanism in analogy to the reaction you know for alkenes. Remember that the main product derives from the rate limiting step having lower activation energy (i.e. proceeding via most stable intermediates/transition states).

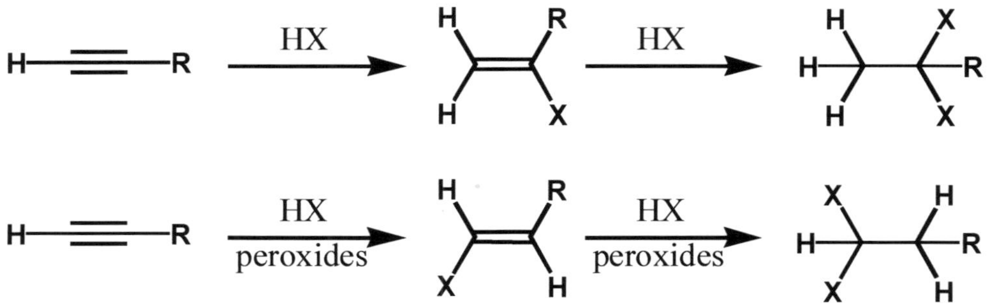

**Notes:**

# Mechanism of Alkyne Hydrohalogenation

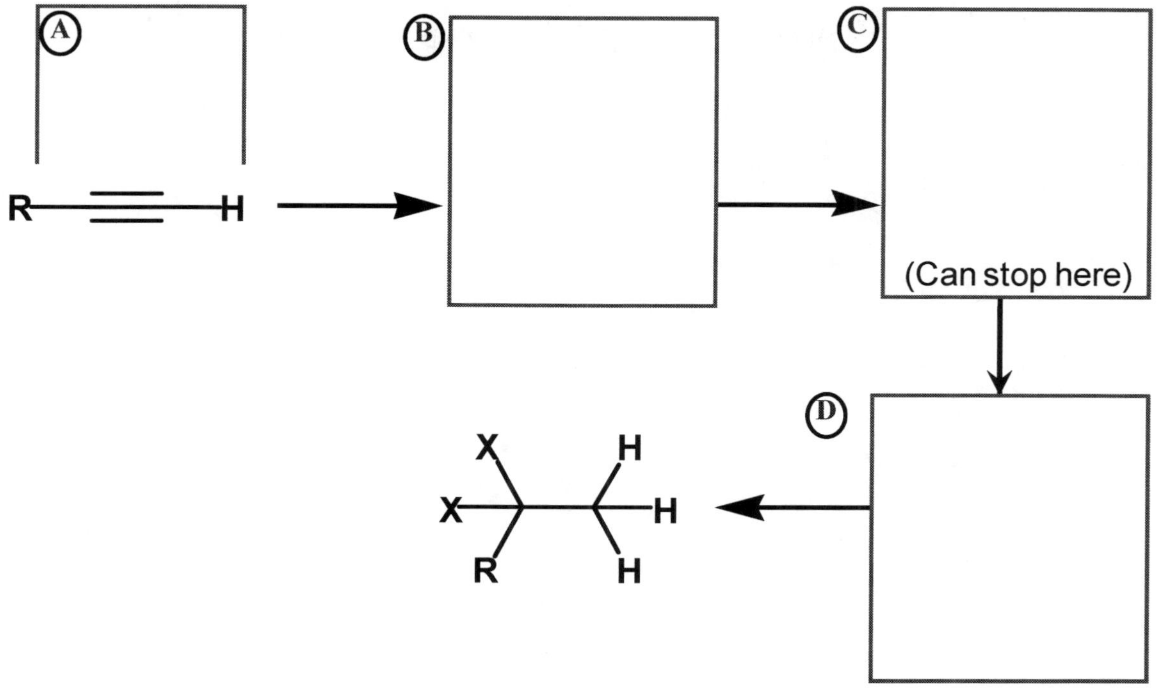


**Notes:**

# Halogenation

Alkynes react with halogens in analogy to alkenes. Appropriate selection of conditions can allow one or two halogen additions across the unsaturated unit:

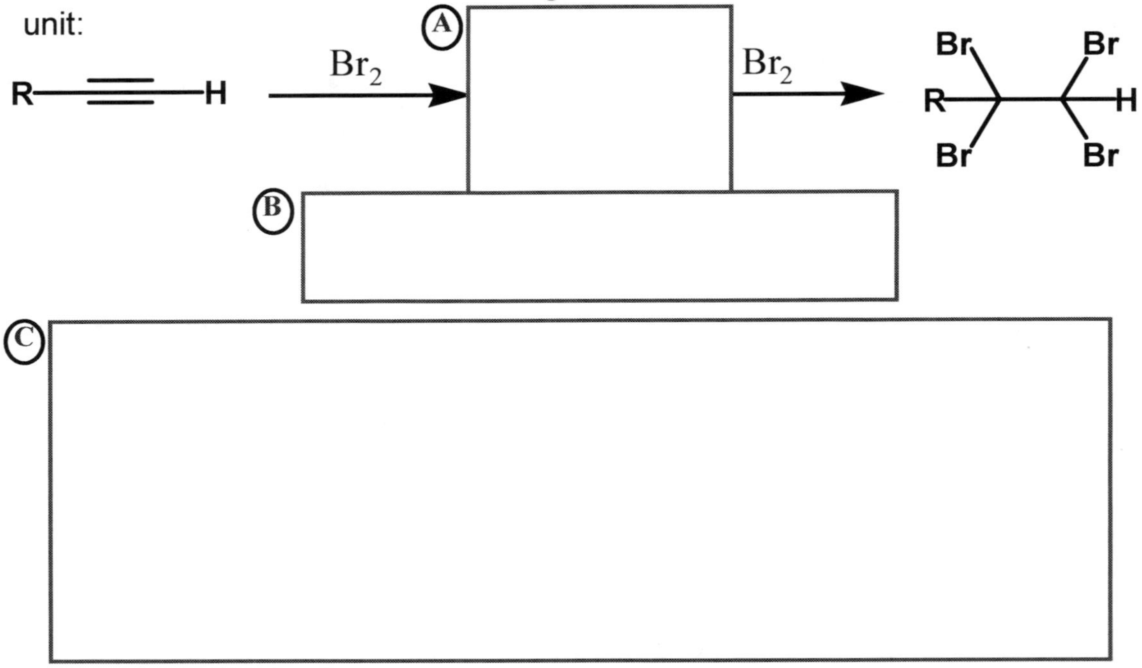


**Notes:**

# Halogenation Example

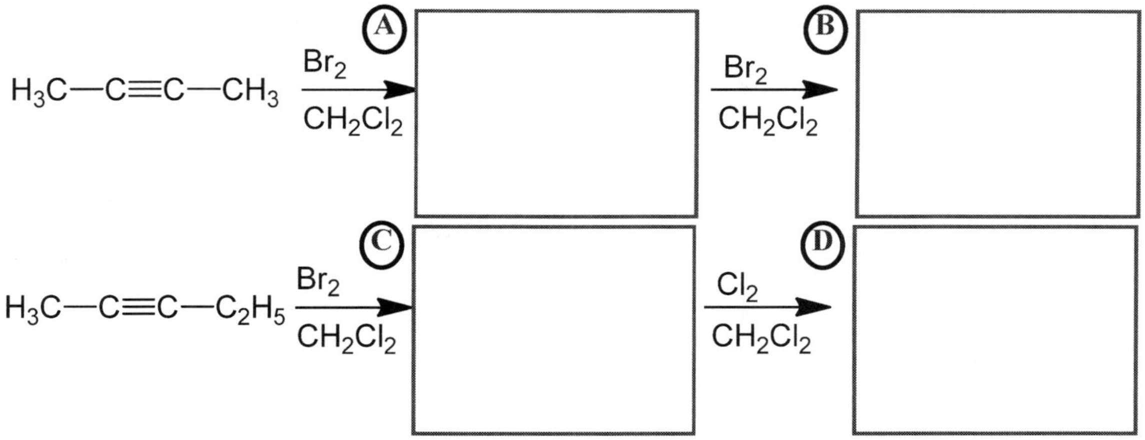


**Notes:**

# Hydration of Alkynes makes <u>KETONES</u>

Alkynes can also undergo hydration reactions:

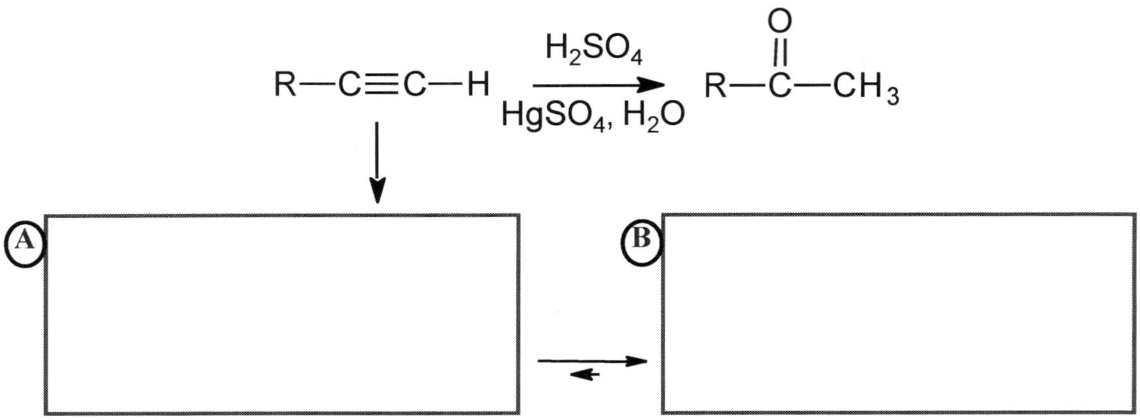


Markovnikov!

Tautomerization

C

**Notes:**

# Constitutional Isomers: Tautomerization

Tautomers:

A

B

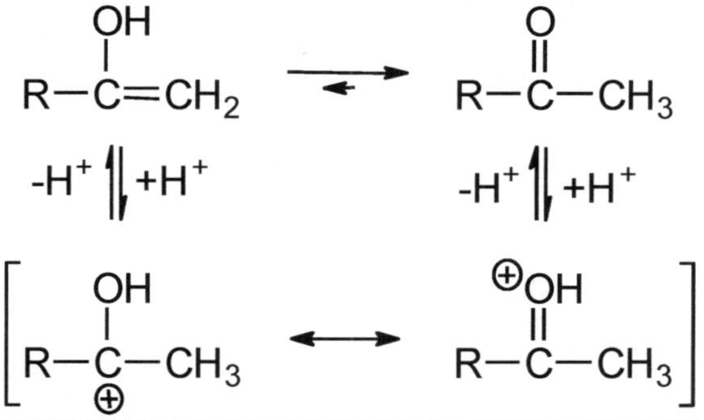


C

D

This is a VERY important concept, which you should learn about extensively in Organic 2.

**Notes:**

# Hydration Problem

Example: Give the products.

$$CH_3-C\equiv C-H \xrightarrow[H_2SO_4]{HgSO_4}$$

$$CH_3-C\equiv C-CH_3 \xrightarrow[H_2SO_4]{HgSO_4}$$

$$CH_3CH_2-C\equiv C-CH_3 \xrightarrow[H_2SO_4]{HgSO_4}$$

**Notes:**

# Hydroboration: Anti-Markovnikov

Internal: KETONES

(A)

Terminal: ALDEHYDES

(B)

Only one alkyne could form an aldehyde via hydration; many can via hydroboration

**Notes:**

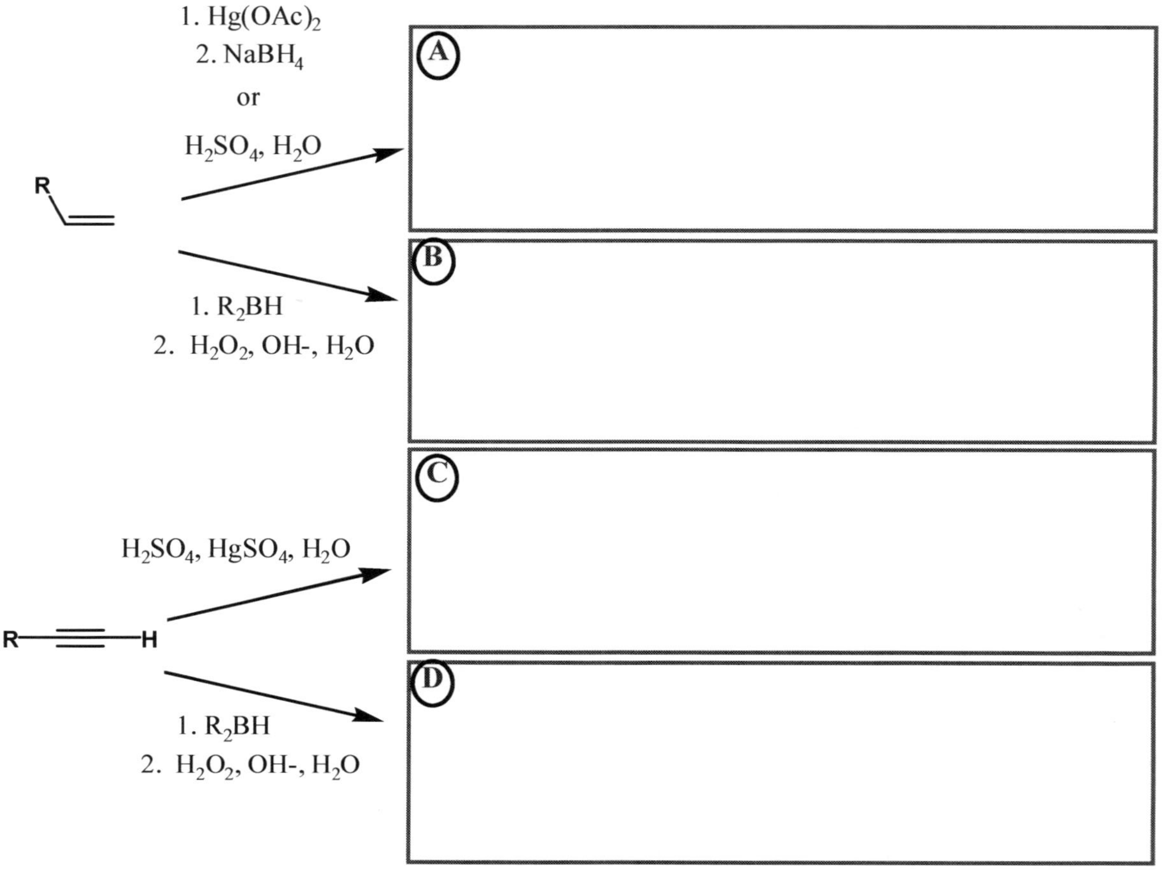
1. Hg(OAc)$_2$
2. NaBH$_4$
or
H$_2$SO$_4$, H$_2$O
R
A
B
1. R$_2$BH
2. H$_2$O$_2$, OH-, H$_2$O
C
H$_2$SO$_4$, HgSO$_4$, H$_2$O
R
H
D
1. R$_2$BH
2. H$_2$O$_2$, OH-, H$_2$O

**Notes:**

# Acetylide Anions and Preparation of Alkynes

A **strong** base is needed to prepare acetylide anions:

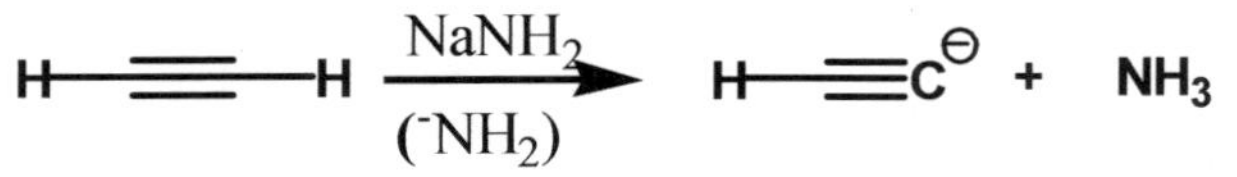

Base = $OH^-$; NO!

(A)

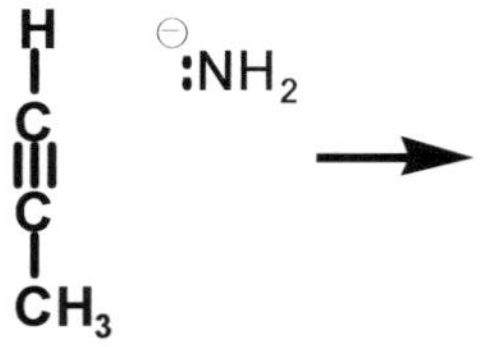


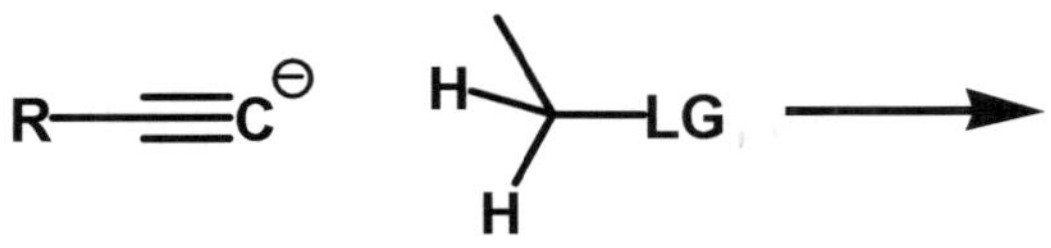


LG = "leaving group"

**Notes:**

# Reaction of Acetylides with Alkyl Halides

An example is the reaction of acetylide anions with PRIMARY alkyl halides, resulting in the alkylation of the acetylide via an

(A)

The reaction can be carried out consecutively to make asymmetrically substituted alkynes:

$$\text{H}-\text{C}\equiv\text{C}-\text{H} \xrightarrow{\text{NaNH}_2} \text{H}-\text{C}\equiv\text{C}^{\ominus} \xrightarrow[\text{RX}]{1^\circ \text{ alkyl halide}} \text{H}-\text{C}\equiv\text{C}-\text{R}$$

(B)

(C)

Here, our leaving group = halide ($X^-$)

**Notes:**

# Preparation of Alkynes

Alkynes can be prepared by sequential elimination of two equiv HX from an alkyl dihalide via an E2 mechanism, in a manner analogous to that used to prepare alkenes:

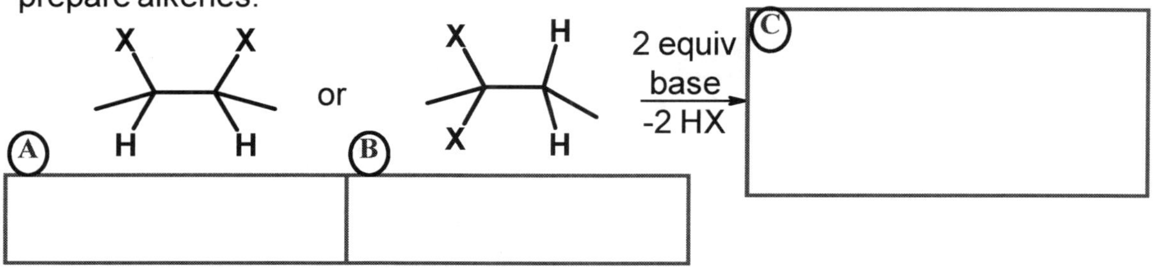


The second elimination is far more difficult, and the reaction can be stopped after one elimination:

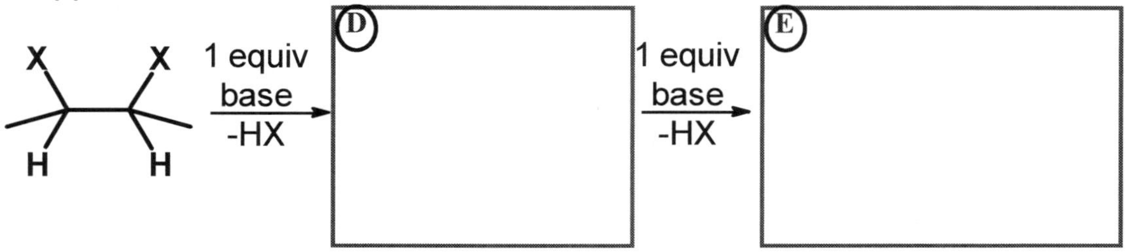


Base = KO$^t$Bu or $NaNH_2$, for example

**Notes:**

# Lecture Set 8: Nucleophilic Substitution Reactions ($S_N1$ and $S_N2$) of Alkyl Halides

**Suggested Reading:**

**Suggested Problems:**

# Nucleophilicity versus Basicity

Nucleophilicity is a measure of how readily a compound (a nucleophile) is able to attack an electron-deficient atom

**Nucleophilicity is measured by a rate constant ($k$).**

(A)

**UNLIKE**

Basicity, which is a measure of how favorable it is for a compound (a base) to shares its lone pair, often with a proton, in solution.

**Basicity is measured by the acid dissociation constant ($K_a$).**

(B)

**Notes:**

# Nucleophilicity versus Basicity

**Notes:**

# Reaction of ROH with HX

Alcohols can react with HX to form alkyl halides. The net reaction is:

$$\textbf{R-OH} \quad + \quad \textbf{H-X} \longrightarrow \textbf{R-X} \quad + \quad \textbf{H-OH}$$

This is a **substitution reaction**.

This reaction proceeds by a specific mechanism called the $\mathbf{S_N1}$ mechanism. This is an abbreviation meant to help you remember how the reaction takes place:

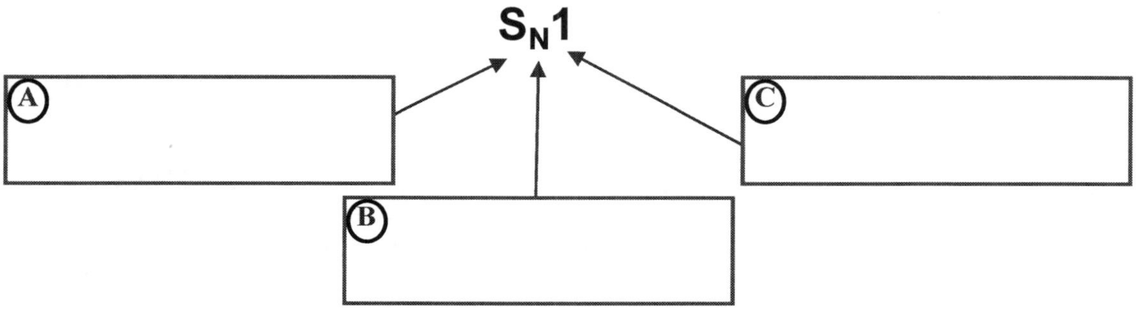


How does this reaction take place? What is the arrow-pushing **mechanism**?

**Notes:**

# Substitution Reaction: $S_N1$

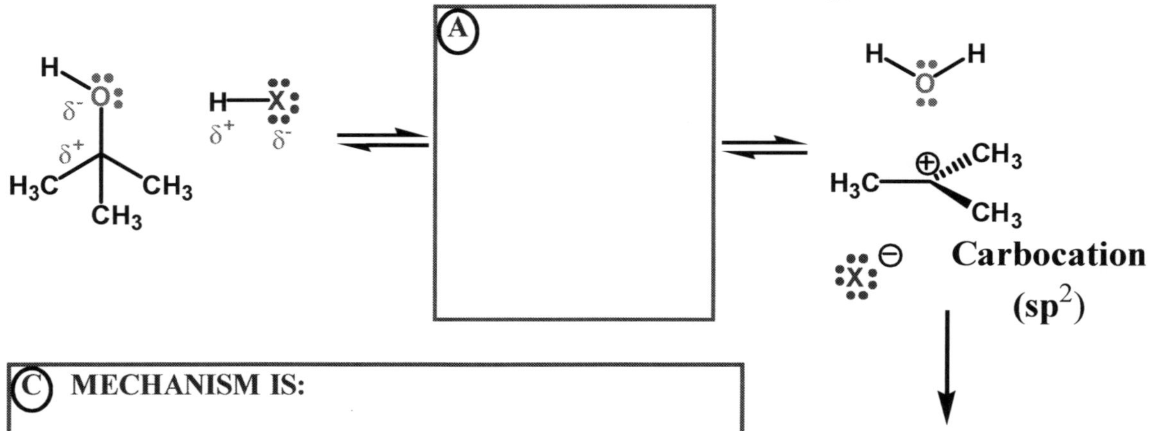

C MECHANISM IS:

Cation can:

1. Activate leaving group
2. Group leaves
3. Anion substitutes

**Notes:**

# Stability of Carbocations

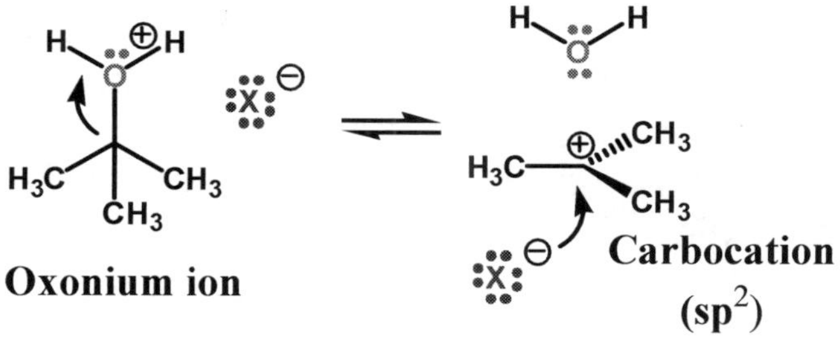


Remember that the rate limiting step is cation formation, and forming the more stable cation is faster. This is why we observe the trend in alcohol reactivity: (A)

REACTIVITY

**In fact: (B)

**Notes:**

The larger the area over which electrons or charge can be effectively delocalized, the greater the resonance stabilization energy. Consider a simple hydrocarbon in which we have one C-C double bond and one C-C single bond. An anion, a cation, or a radical (placed at the carbon drawn as 1° $sp^3$-hybridized in resonance contributors) all have a special stability versus the same species at typical primary $sp^3$-hybridized carbons because the electrons/charge can be delocalized over the molecule as demonstrated by the resonance contributors on the left and the resonance hybrids (actual structures) on the right.

**Resonance Contributors** | **Resonance Hybrid**

**Allyl Radical**

**Allyl Cation**

**Allyl Anion**

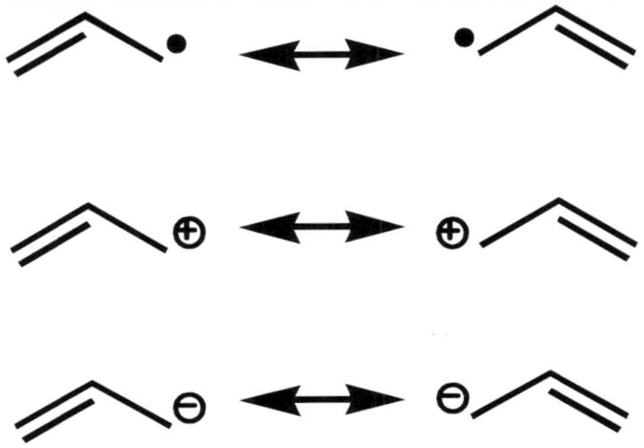

**Notes:**

# The Benzylic Position

An sp³-hybridized carbon adjacent to a phenyl ring is called a

Ⓐ

Because reactive species formed at the benzylic position are stabilized by delocalization into the pi system, such systems display special reactivity.

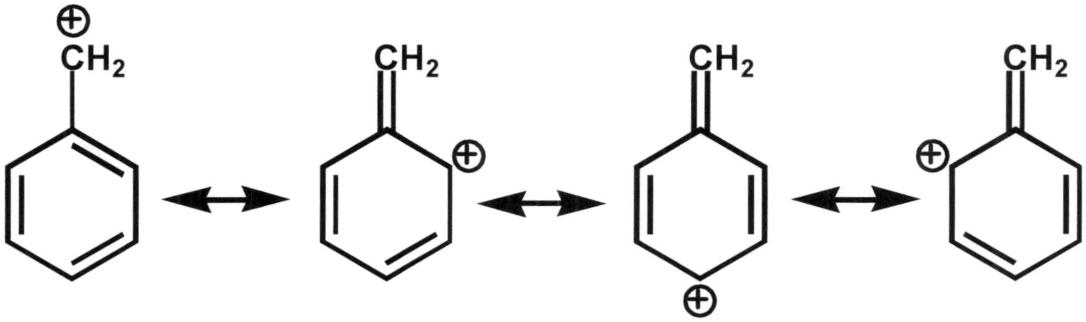


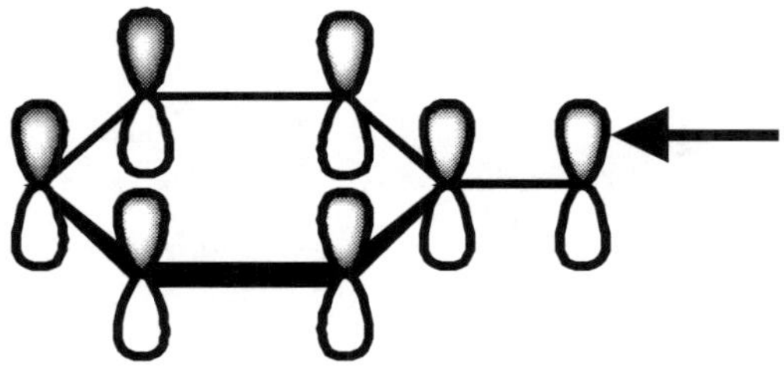

Ⓐ

**Notes:**

# Relative Stability of Carbocations

Increasing
stability of cations

**Notes:**

# Relative Stability of Radicals

Increasing stability
of radicals

**Notes:**

# Allylic Cations: $S_N1$

$S_N1$ is facilitated at an allylic site by resonance stabilization of the carbocation intermediate:

**Relative Rate**

1

125

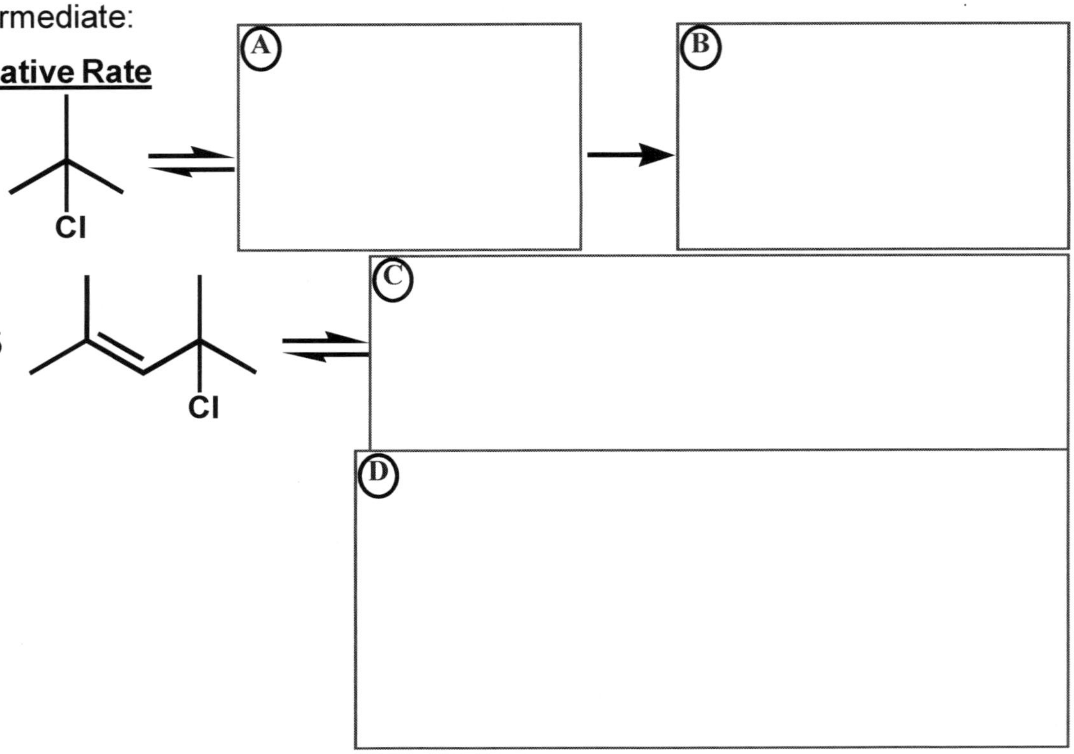


**Notes:**

# Allylic Cations: $S_N1$

But keep in mind that the **major product** formally results from the **more stable resonance contributor** of the carbocation intermediate; if the system is asymmetric, can get two products:

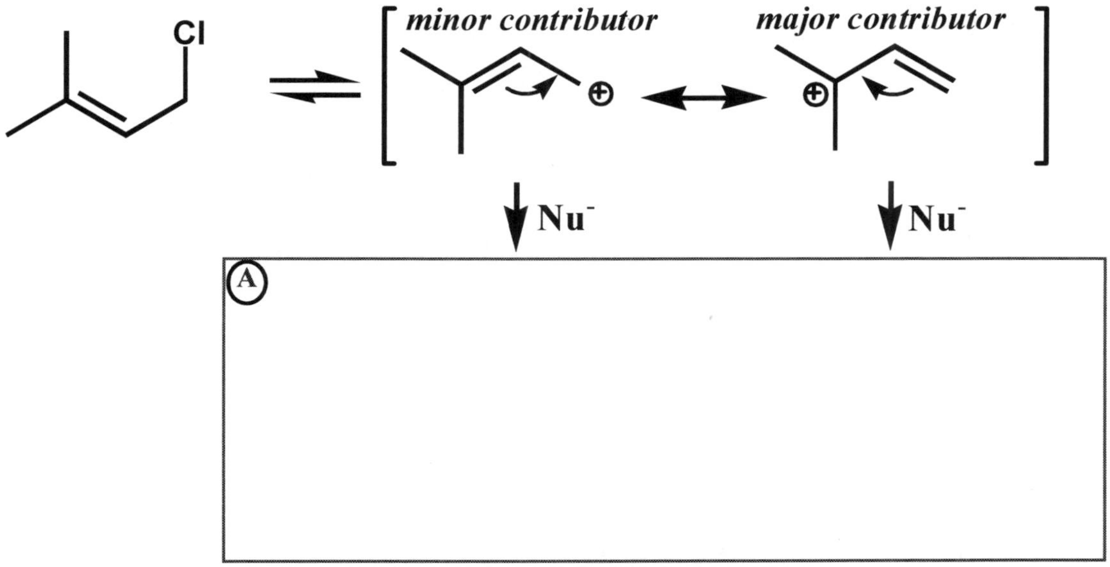


**Notes:**

# Allylic Cations: $S_N2$

$S_N2$ is facilitated by overlap with the adjacent pi orbitals in the transition state:

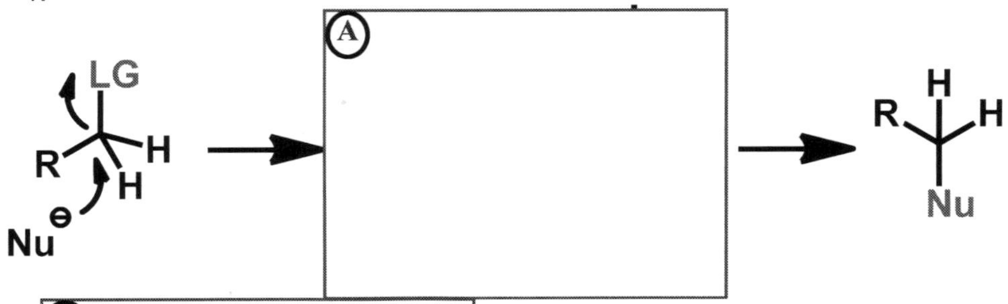


B

typical transition state orbital picture

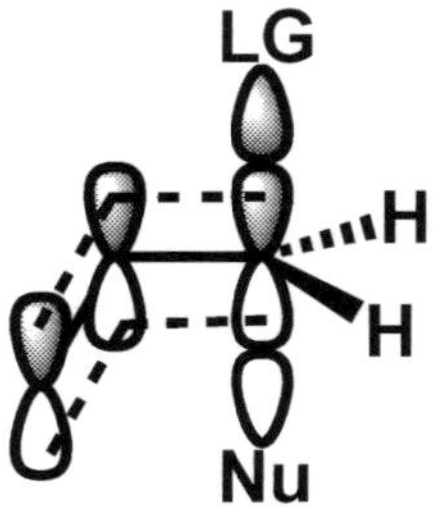


transition state orbital picture where reaction center is an allylic position

**Notes:**

# Benzylic Position: $S_N1$ and $S_N2$

The $S_N2$ and $S_N1$ reactions are facilitated at benzylic positions in the same manner as discussed for allylic systems.

$S_N2$:

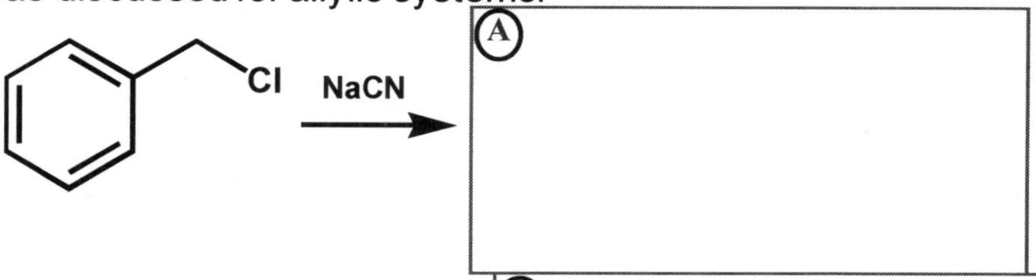


$S_N1$:

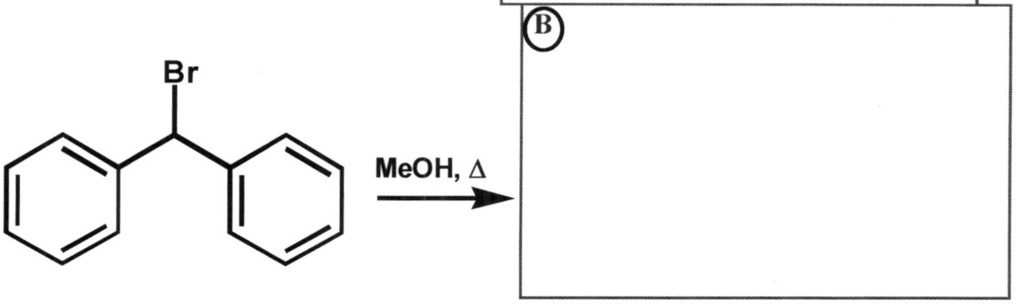


In the $S_N1$ example, reaction is facilitated to an even greater extent by formation of a highly stabilized benzyl cation in the rate limiting step (in this case it can be delocalized into **two** phenyl rings!)

**Notes:**

# The $S_N2$ Reaction Mechanism

Because (A)

reaction between *primary* alcohol and HX proceeds by a mechanism

(B)

**Primary alcohols** undergo the

(C)

The reactants and products can be the same as in the $S_N1$ case, but the MECHANISM is different!

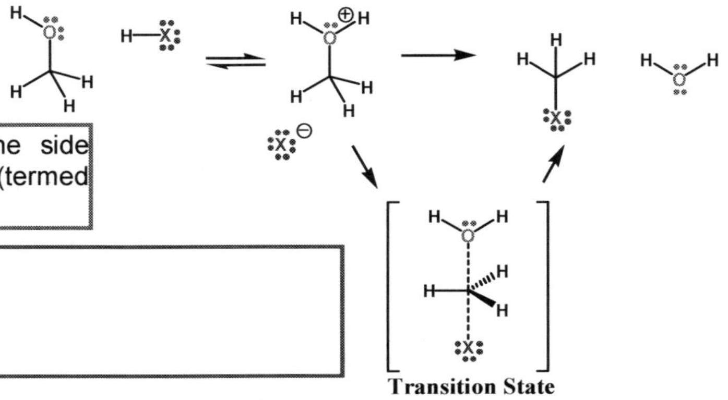


Nucleophile attacks on the side opposite the leaving group (termed "backside attack")

(D)

(recall: dashed line is a *partial* bond, i.e., less than two electrons, so C does NOT have 10 electrons here!

**Notes:**

# Nucleophilic Substitution Reactions

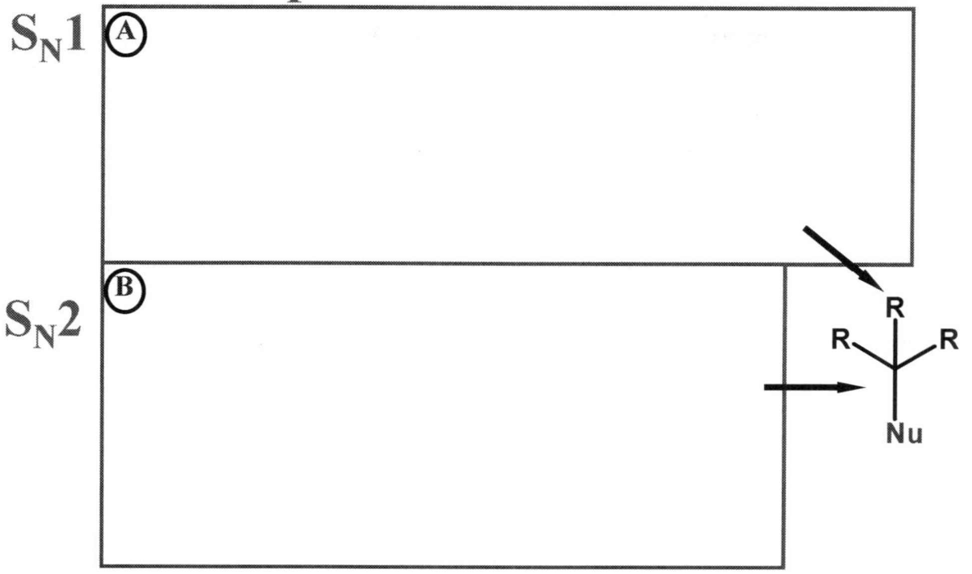


We know the mechanism for $S_N1$ and $S_N2$ reactions. Because both reactions work on the same types of starting materials to yield the same types of products, how do we know which reaction is taking place, and does it matter? (LG = leaving group, Nu = nucleophile)

**Notes:**

# $S_N1$ versus $S_N2$

$S_N1$ reaction is favored for substitution on a tertiary carbon.

| (A) |
|---|

$S_N2$ is favored for substitution on a primary carbon.

| (B) |
|---|

The two mechanisms can sometimes compete with one another. The reaction mechanism which predominates in a particular case depends on :

| |
|---|
| (C) |
| (D) |
| (E) |
| (F) |

**Notes:**

# Need a good leaving group for $S_N1$ or $S_N2$

Note that in either $S_N1$ or $S_N2$ reaction of an alcohol, the OH functionality must be activated (i.e., by protonation) so that when it is substituted by $X^-$ it will form a stable molecule (i.e. water):

Ⓐ

In fact, under basic conditions, an alkyl halide can be converted to an alcohol (the opposite of the previous reaction!), by $S_N1$ or $S_N2$ routes:

Ⓑ

Conditions dictate products!

**Notes:**

# <u>Four Factors</u> Affecting $S_N2$ Rate

RX and the nucleophile (Nu; $OH^-$ in this example) come together to form the transitions state in the rate-limiting step.

The reaction will be faster if the energy of activation ($\Delta G^{\ddagger}$ in the diagram) is lowered. What factors influence the energy of the transition state?

| |
|---|
| (A) |
| (B) |
| (C) |
| (D) |

**Notes:**

# $S_N2$: Leaving Group Effects

It is energetically unfavorable to produce high-energy (reactive) species; a reaction is favored when lower energy (more stable) species are produced; MORE STABLE ANION = BETTER LEAVING GROUP!

reactivity: > > >>

best LG worst LG

**Relative Rates of Reactions**

**30,000**

**10,000**

**200**

**1**

**Remember the trends, not the numbers

**Notes:**

# $S_N2$ : Steric Effects (RX)

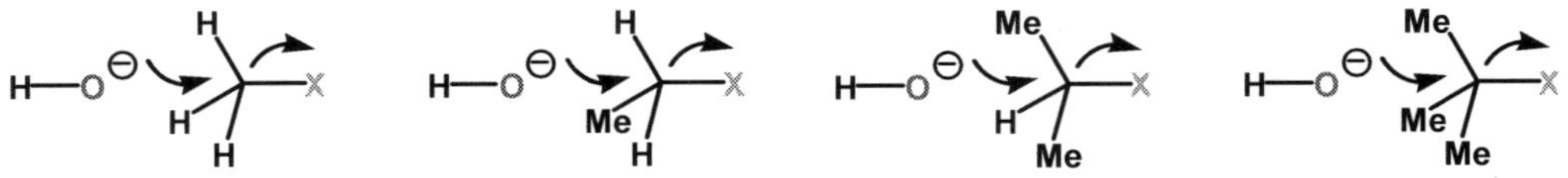


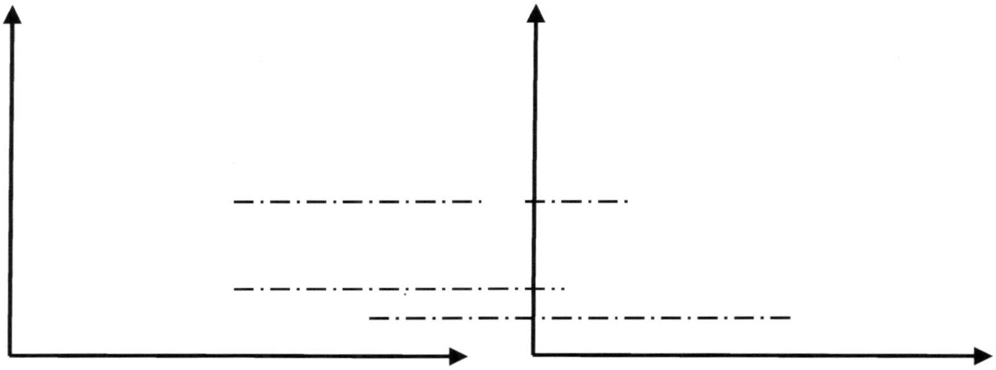

(A)

**Notes:**

# $S_N2$: Steric Effect (Nucleophile)

We have seen that a bulky alkyl halide reacts more slowly than a less bulky one. Bulky substituents also diminish the reactivity of a potential nucleophile. Ethoxide is a better nucleophile than *t*-butoxide, for example.

(A)

(B)

(C)

**Notes:**

# The Solvent Influences the Nucleophilicity

A

B

It is easier to break the ion-dipole interactions between a weak base and the solvent than between a strong base and the solvent

C

**Notes:**

# The Solvent Influences the Nucleophilicity

Aprotic polar solvents such as DMSO facilitate the reaction of ionic compounds because they solvate ions very well. Examples of different classes of solvents:

nonpolar: hexane

moderately polar: diethylether, acetone

polar protic: $H_2O$, ROH

polar aprotic: DMSO
(dimethylsulfoxide)

DMF
(*N*.*N*-dimethylformamide)

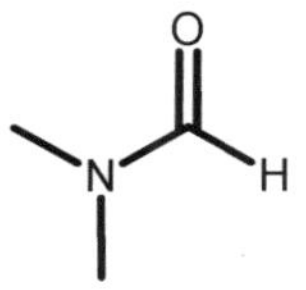


**Notes:**

# Solvent Effects

(A)

| | | |
|---|---|---|
| $OH^-$ | in $H_2O$: | strong ion-dipole interaction with water diminishes its nucleophilicity (less reactive; higher $E_a$) |
| | in DMSO: | weaker solvation and thus more nucleophilic (faster reaction) versus the same reaction in water |

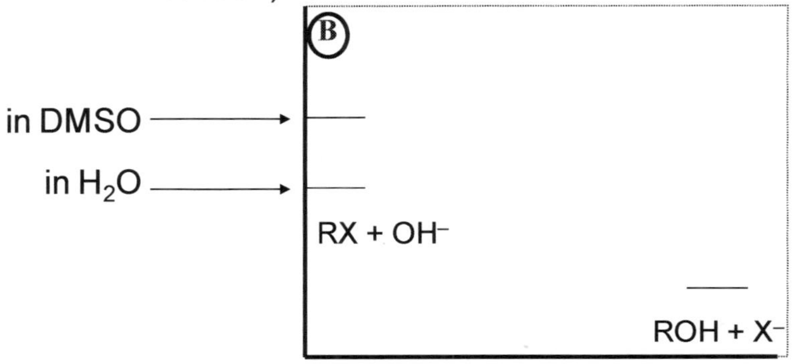


**Notes:**

# Nucleophile Trends

## Nucleophilicity Trends

Ⓐ

$O^{2-} > HO^- > H_2O$
$RO^- > ROH$
$HS^- > H_2S$

Ⓑ

$H_2N^- > HO^- > F^-$
$H_3N > H_2O$
$RO^- > RCO_2^-$

Ⓒ

$I^- > Br^- > Cl^- > F^-$
$HS^- > HO^-$
$PH_3 > NH_3$

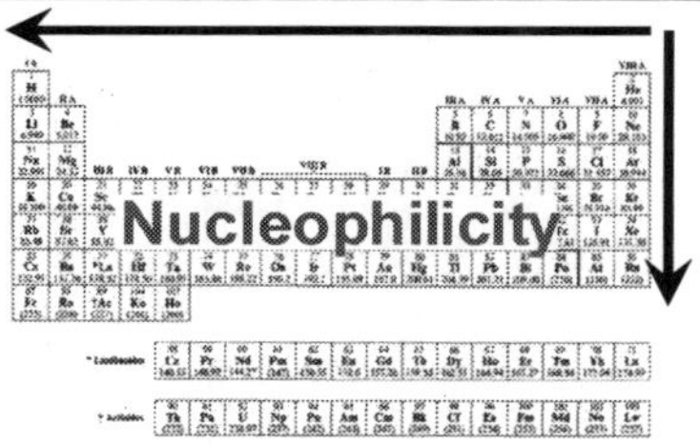


**Notes:**

# $S_N2$ Mechanism Problem

Example. Which reaction will proceed faster in each of the following pairs?

(A)

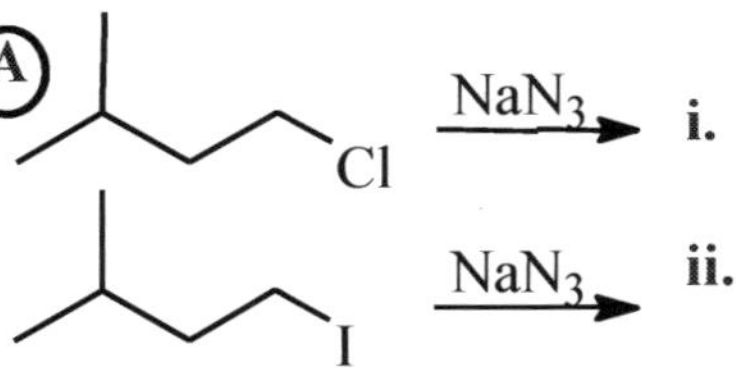


(B)

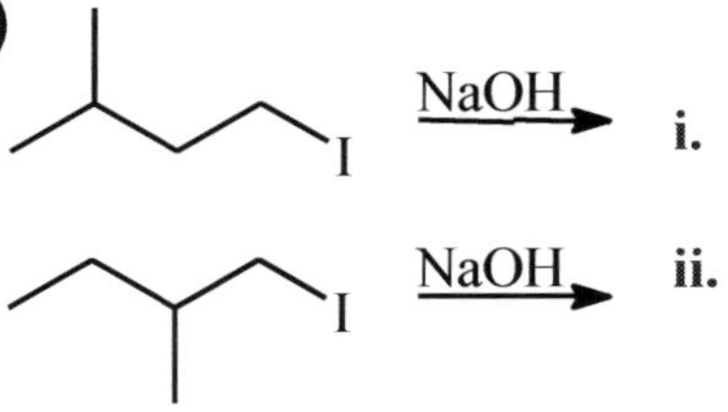


(C)

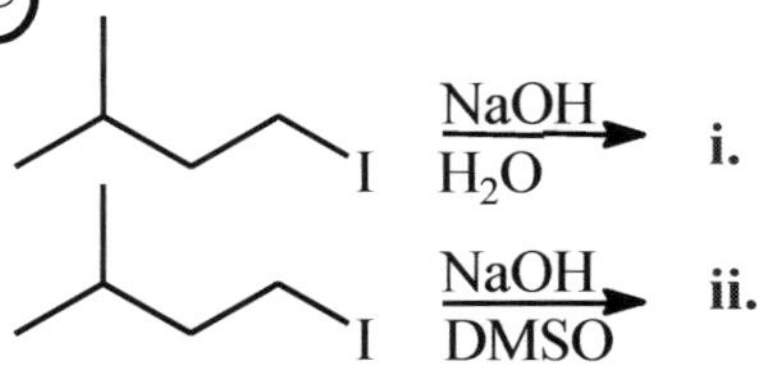


(D)

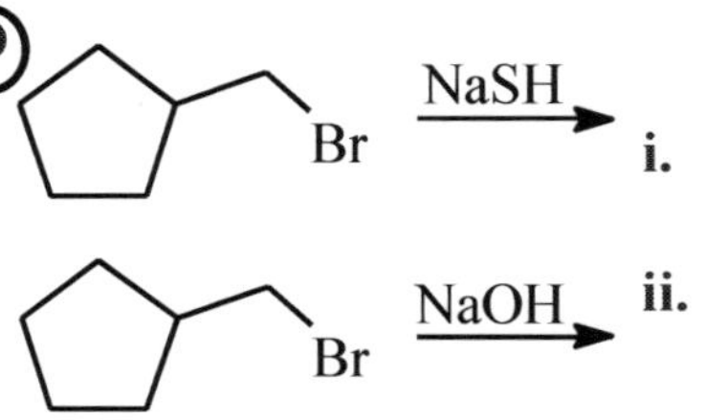


**Notes:**

# $S_N1$ mechanism: compare to $S_N2$

Kinetics:

Rate = $k$[RX]

(a first-order reaction)

| |
|---|
| Rate-limiting step is:<br>(A) |
| Therefore:<br>(B) |

**Notes:**

# $S_N1$ Mechanism: Leaving Group

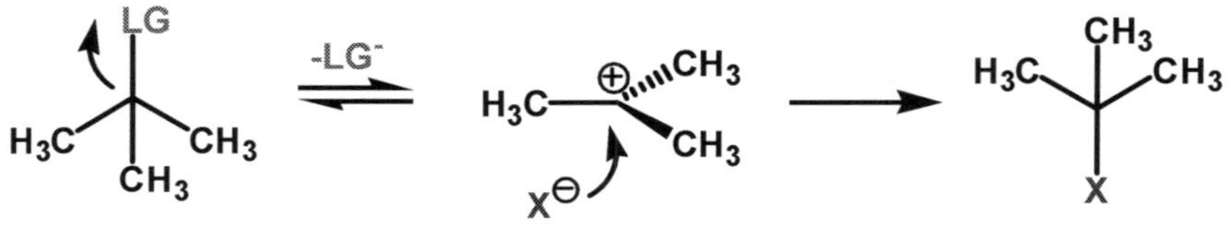


Notice that the leaving group (LG) leaves in the rate limiting step; thus,

(A)

However, the nucleophile

(B)

**Notes:**

# $S_N1$ Solvent Effects

nonpolar: (A)

moderately polar: (B)

polar protic: (C)

polar aprotic: (D)

The $S_N1$ mechanism is promoted by **polar protic solvents**, because

(E)

(F)

(G)

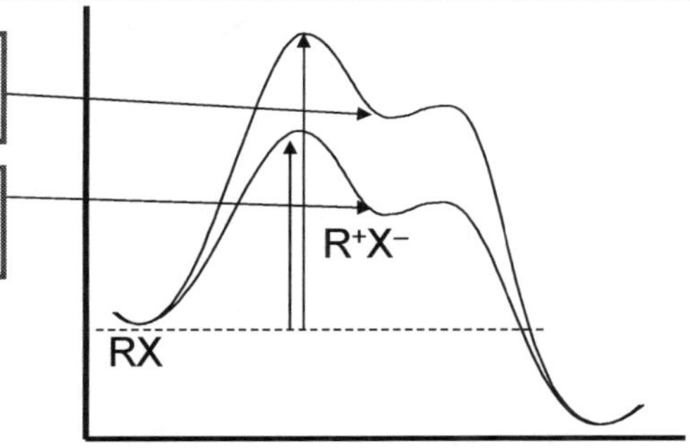


**Notes:**

# Stereochemistry: $S_N2$ is Stereospeecific

The back side attack in the $S_N2$ reaction leads to

(A)

(B)

The $S_N2$ reaction is an example of a **stereospecific** reaction:

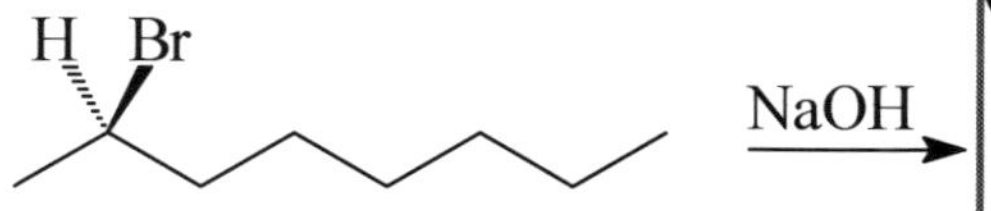


(*R*)-2-bromooctane

(C)

(*S*)-2-octanol

**Notes:**

# Stereochemistry: $S_N1$ is Stereorandom

(A)

**Notes:**

# Good Nucleophiles

Additional functional groups can be prepared using alkyl halides via a nucleophilic substitution mechanism. The factors affecting substitution reactions apply to all of these!

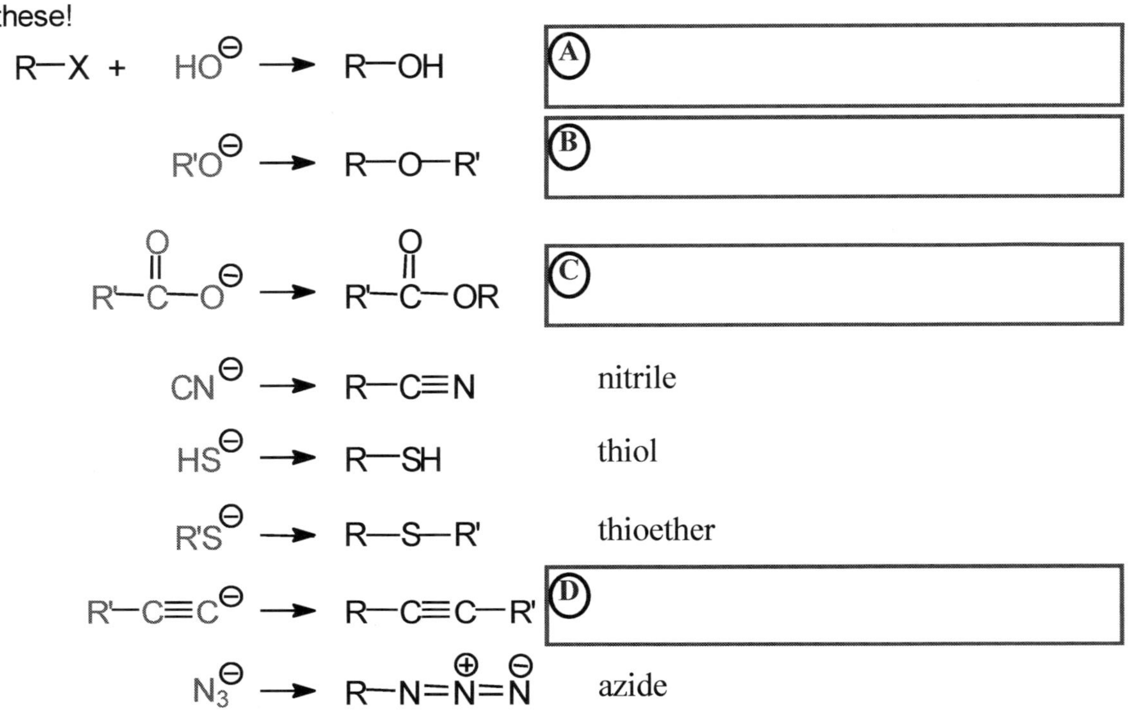


**Notes:**

# $S_N1$ and $S_N2$ Summary

rate of $S_N1$ increases → (more stable carbocation)

RX = $CH_3X$ 1° 2° 3°

← rate of $S_N2$ increases (more steric hindrance)

$S_N2$

(A)

$S_N1$

(B)

**Notes:**

# Lecture Set 9:
# Elimination Reactions (E1 and E2)

**Suggested Reading:**

**Suggested Problems:**

# The Care Which Project

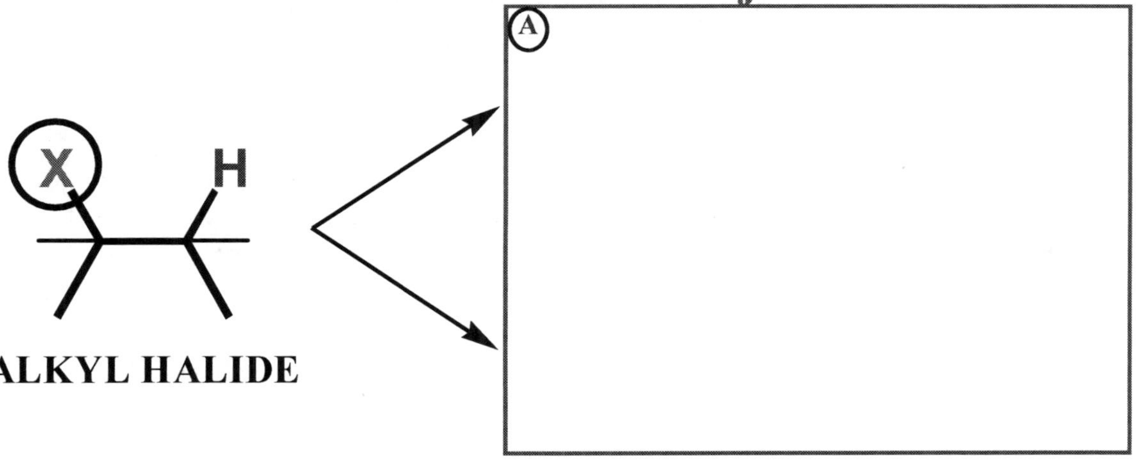


**ALKYL HALIDE**

In a given context: 1. Elimination or Substitution?
2. E1, $S_N1$, E2, or $S_N2$???

How can we tell, and do we care which it is?

LS8: We can anticipate whether $S_N1$ or $S_N2$ predominates.
LS9: Predict competition between $S_N1$, $S_N2$, E1, and E2.

**Notes:**

# Mechanism of Dehydration: E1

These are examples of **E1** reactions (E for elimination, 1 for unimolecular)

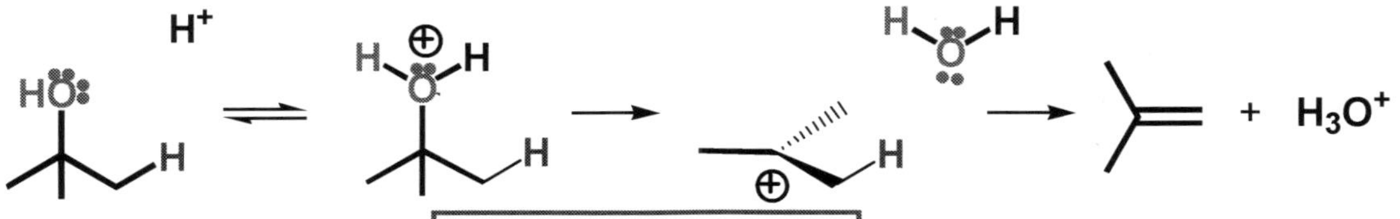


(A)

The **E1** reaction proceeds via carbocation intermediate formation as the rate limiting step.

(B)

**Notes:**

# The E1 Reaction: Dehydrohalogenation

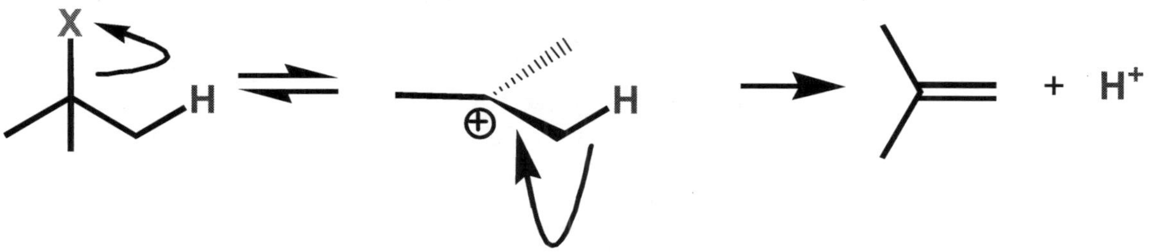


Formation of the carbocation is (A)

The cation can

(B) , leading to multiple potential products.

(C) applies to determining the major product.

**NOTE** (D)

**Notes:**

# E1 Stereochemistry: Most Stable Favored

1. Make the MOST SUBSTITUTED (most stable) alkene.

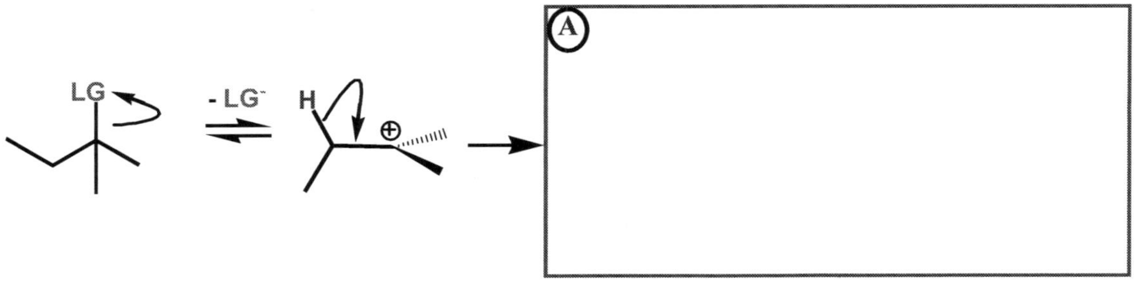


2. If the most substituted choice has more than one configurational isomer, choose the less stericallyhindered:

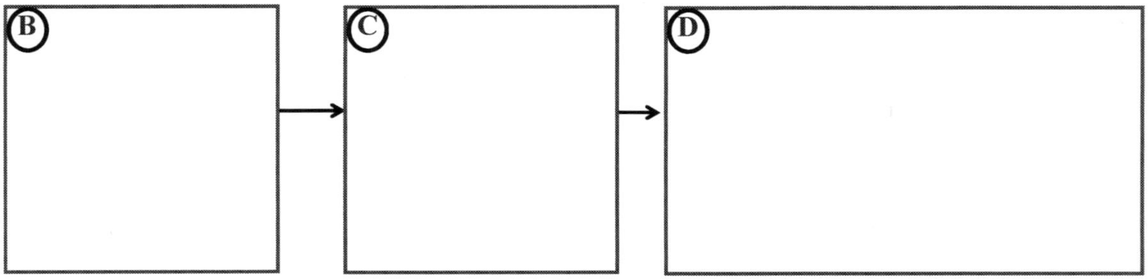


**Notes:**

# E1 Stereochemistry: Most Stable Favored

(A)

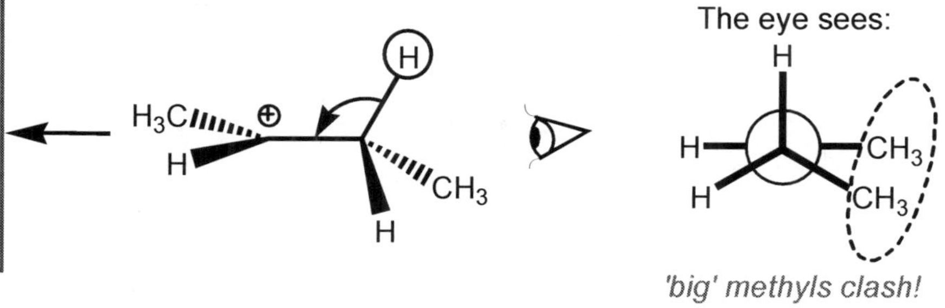


(B)

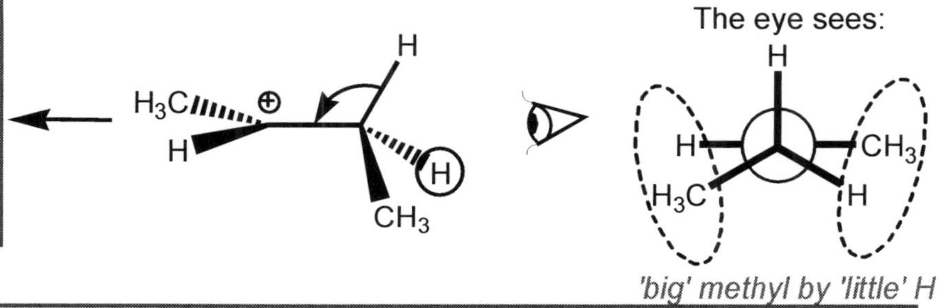


(C)

**Notes:**

# E2 Reaction

Concerted mechanism:

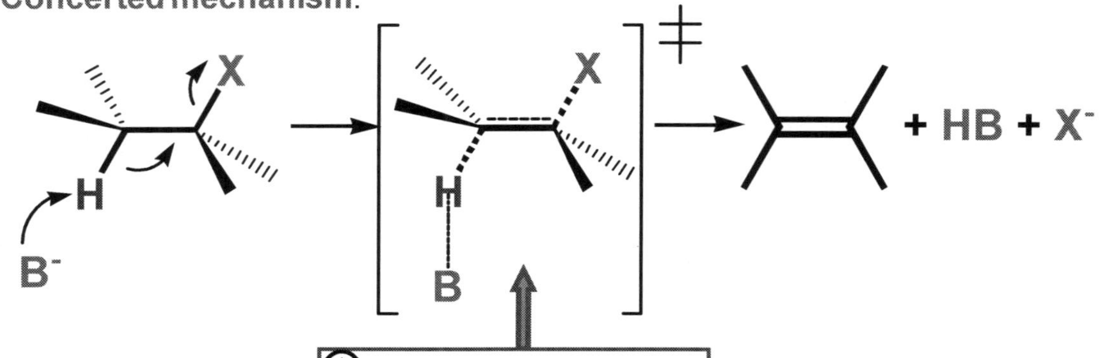


Ⓐ

In order for the E2 reaction to work, the H and X to be eliminated must be

Ⓑ

"Antiperiplanar" means the two groups are *anti-* to one another and in the same plane.

When you have a choice, still want to make the most stable alkene.

**Notes:**

# E2 Reaction: Stereochemistry

There are instances when only one H is present for elimination from a chiral center, leading to the most substituted alkene; in these cases the antiperiplanar requirement may lead to one and only one alkene stereoisomer:

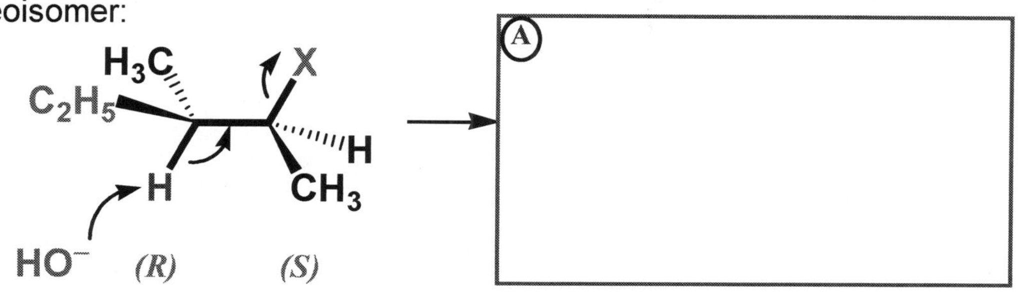


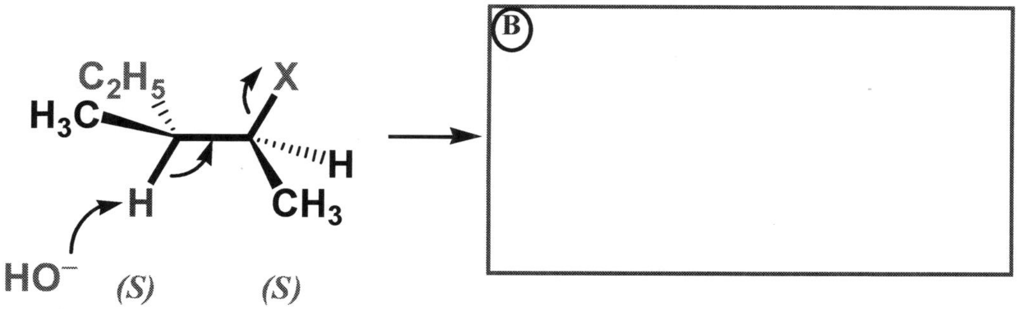


**Notes:**

# E2 Reaction Sterics: non-Zaitsev products

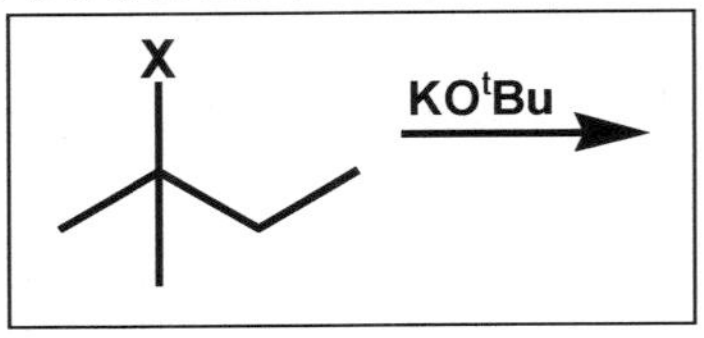


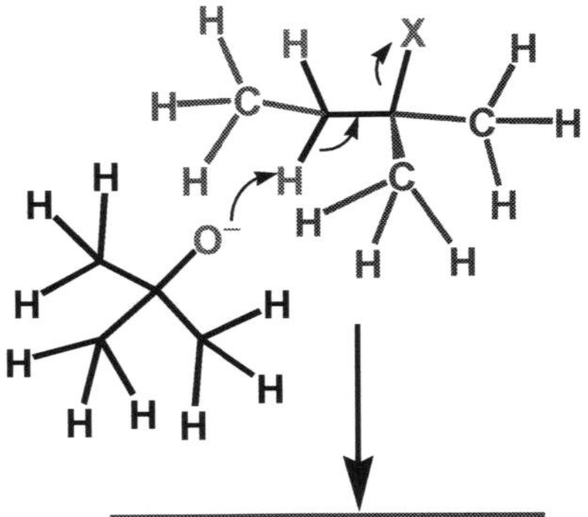

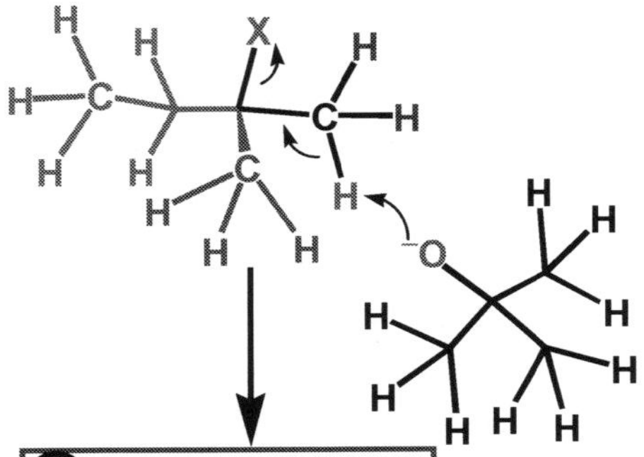

A

B

**Notes:**

# Benzylic Positions: E1 and E2

Eliminations reactions (E1, E2) are also possible at benzylic sites. Like $S_N1$, E1 will be facilitated by stabilization of carbocation intermediates and the alkene that is pi conjugated with the adjacent benzene is favored. E2 will also be facilitated by formation of an alkene that is stabilized by pi conjugation with the benzene system.:

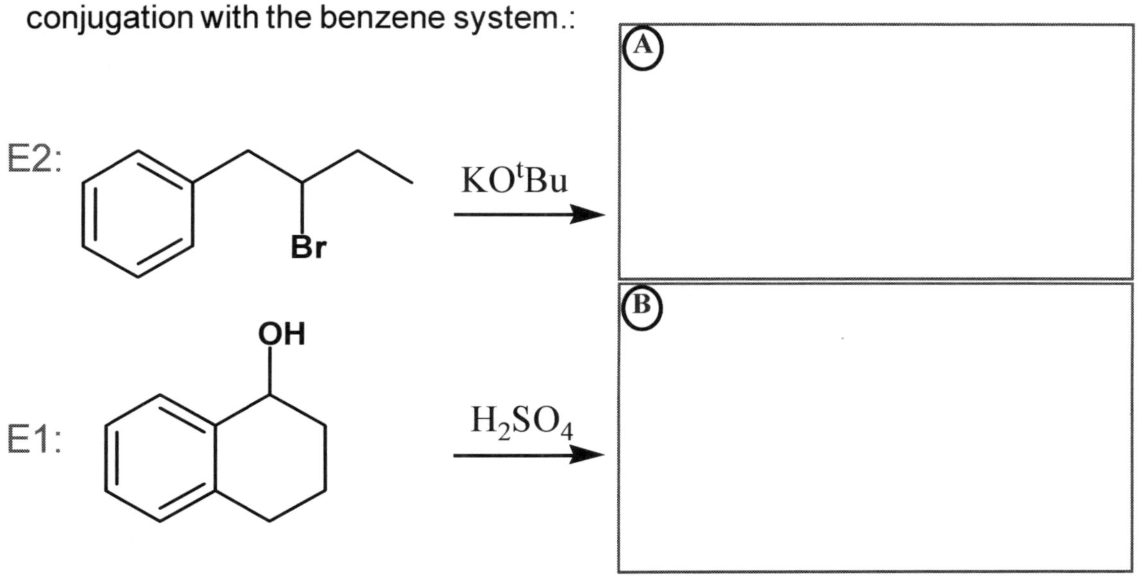


**Notes:**

# The E1 and E2 Reactions: Recap

## E1

1. .

2. .

3. .

4. .

5. .

## E2

1. .

2. .

3. .

4. .

5. .

**Notes:**

# 2 versus 1: Substitution and Elimination

**bimolecular reaction ($S_N2$, E2):**
**favored by high concentration of good Nu or strong B**

**good Nu, weak B** (i.e., $I^-$, $Br^-$, $HS^-$, $NH_3$): (A)

**good Nu, strong B** (i.e., $HO^-$, $EtO^-$, $H_2N^-$): (B)

**poor Nu, strong B** (i.e., $^tBuO^-$, bulky): (C)

Substrate:

**1° RX** no bulky base: (D)

BULKY base (i.e., $^-O^tBu$): (E)

**2° RX** **$S_N2$ and E2**;
the stronger/bulkier the base, the more **E2 predominates**

**3° RX** (F)

**unimolecular reaction ($S_N1$, E1):**
**favored when there is neither a good nucleophile nor a strong Base**

**Notes:**

# Bimolecular Conditions: $S_N2$ versus E2

Base or Nucleophile is Present:

E2 is faster with a stronger (B)

$S_N2$ is faster with a stronger (N)

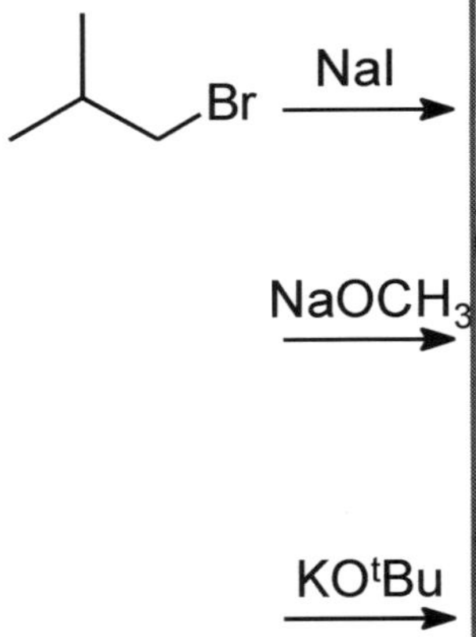


(C)

(D)

(E)

**Notes:**

# Bulky Base Favors E2 over $S_N2$

A bulky base/nucleophile slows down $S_N2$; has less impact on rate of E2
Thus, if you want E2, use a bulky base ($^-O^tBu$)

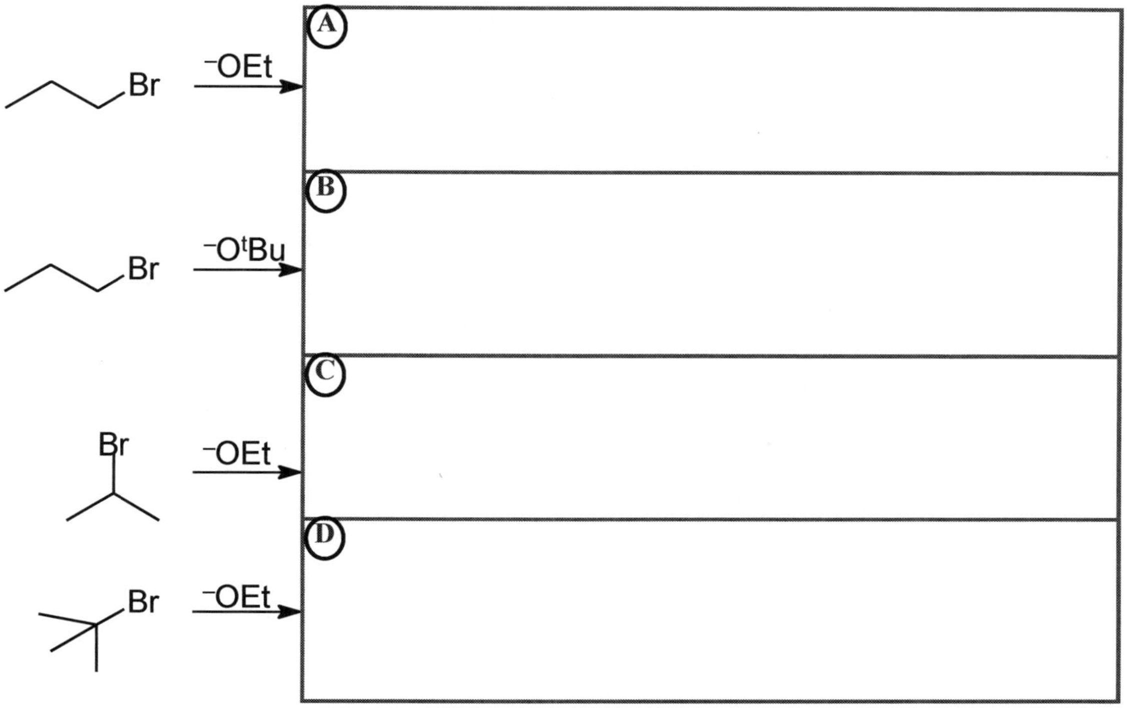


**Notes:**

# Weak Base Favors $S_N2$ over E2

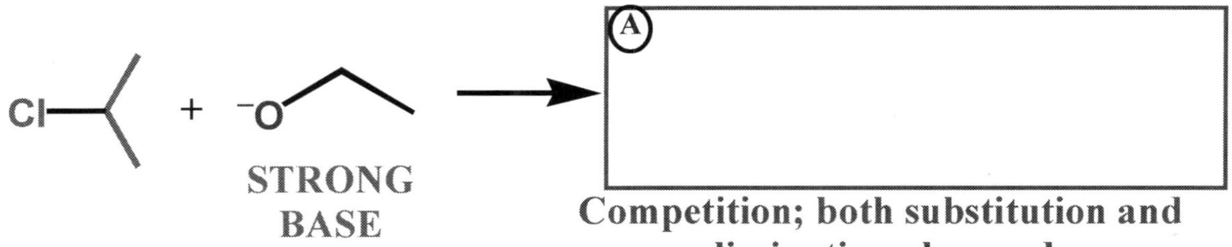


**Competition; both substitution and elimination observed.**

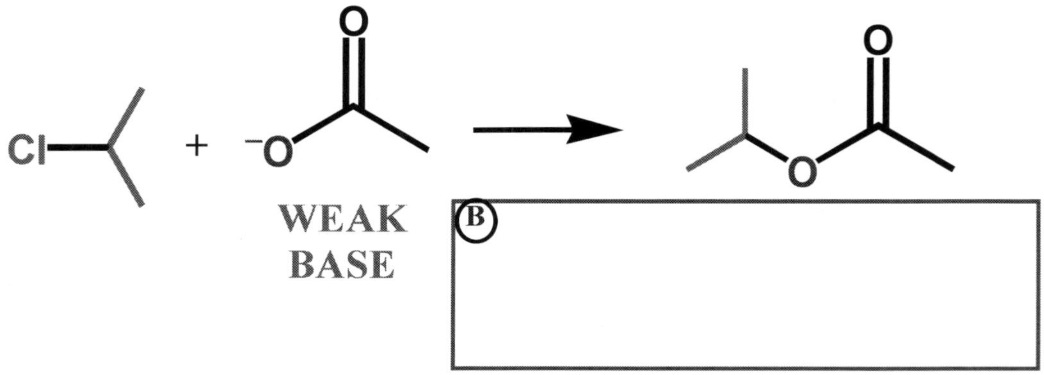


**Notes:**

# $S_N1$ versus E1: Houston, we have a problem!

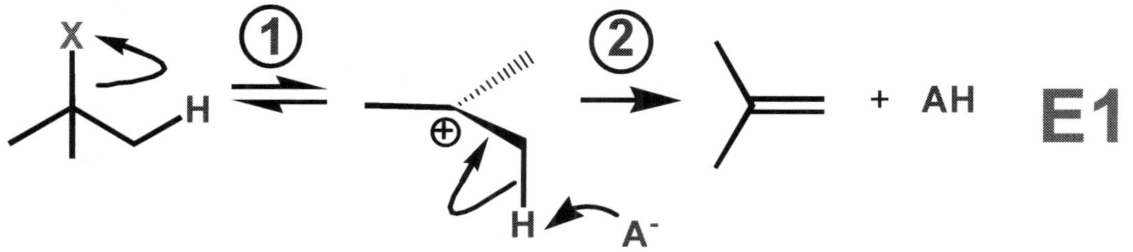


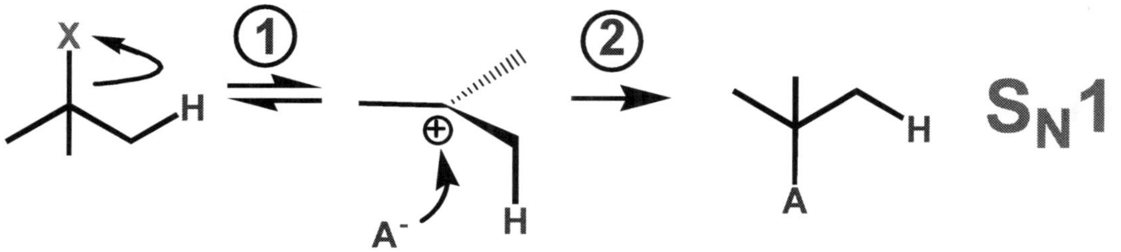


First step: rate = $k[R^+]$

Second step: rate = $k[R^+][A^-]$

(A)

**Notes:**

# Eliminate errors … accept no $S_N$ubstitute

**Notes:**

# Substitution versus Elimination Problem

**Exam**ple. Determine whether each reaction will proceed predominantly via $S_N1$, $S_N2$, E1, E2, or some combination thereof, and show the product(s)

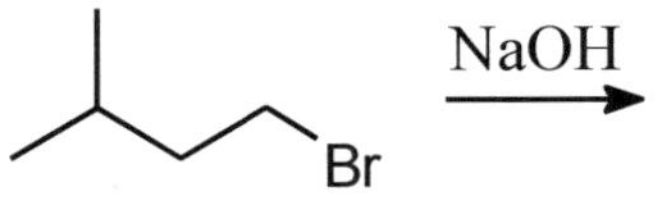


A

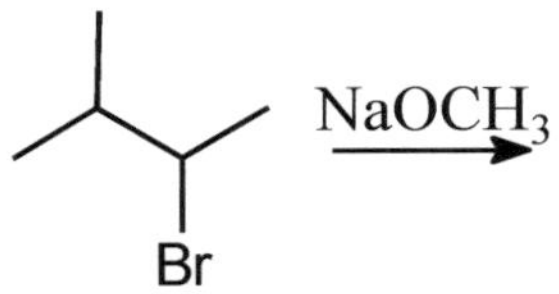


B

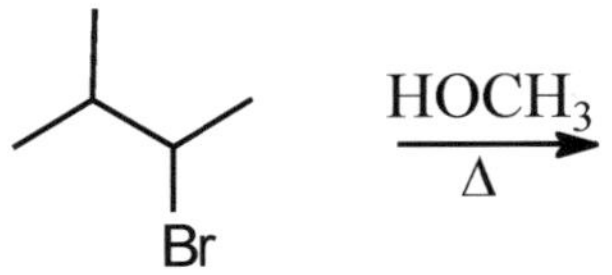


C

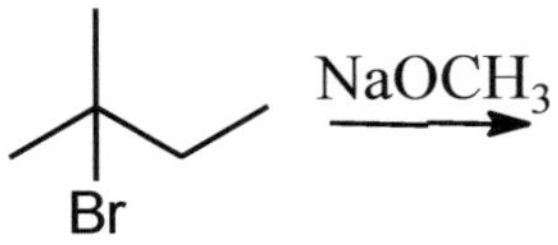


D

**Notes:**

# Reactivity Problem

**Example.** Predict whether each proceeds via $S_N1$, $S_N2$, E1, or E2, and draw the major product (show stereochemistry where applicable).

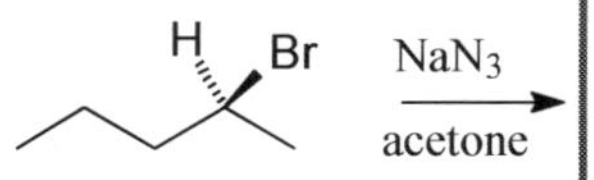


(A)

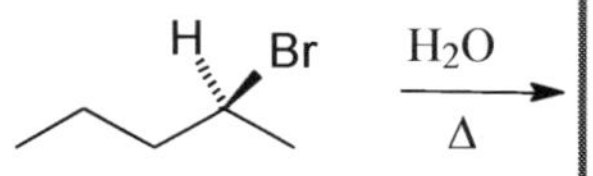


(B)

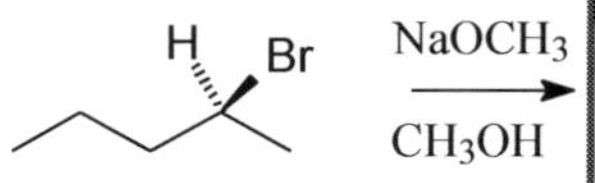


(C)

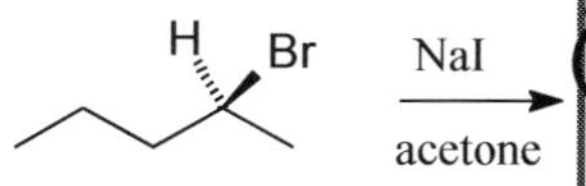


(D)

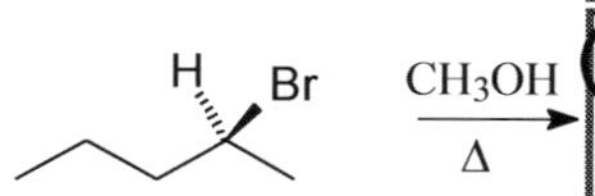


(E)

**Notes:**

# Benzylic Position: $S_N1$, $S_N2$, E1, or E2?

Example. Give the major product(s) and assign each as proceeding predominantly by E1, E2, $S_N1$, or $S_N2$.

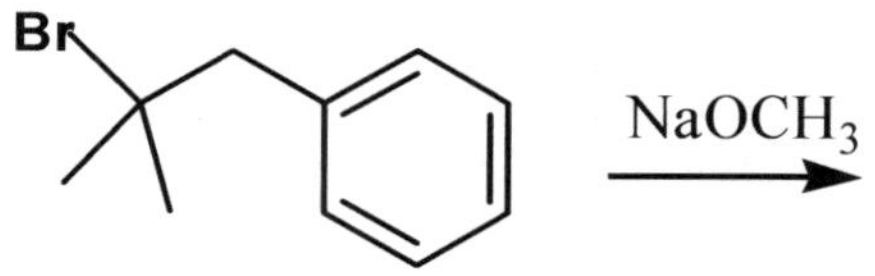

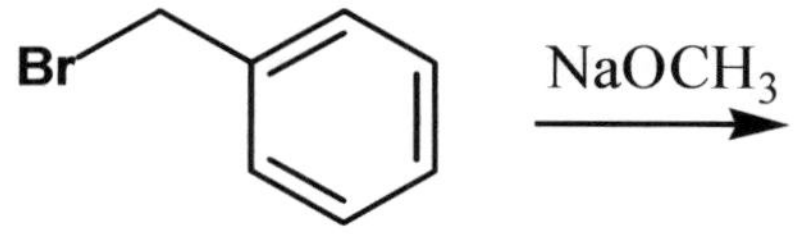

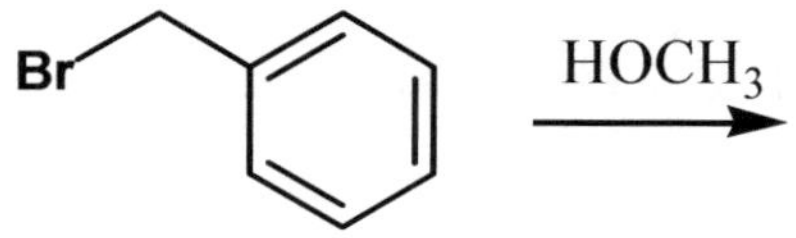



**Notes:**

# Applications: Williamson Ether Synthesis

Williamson Ether Synthesis:

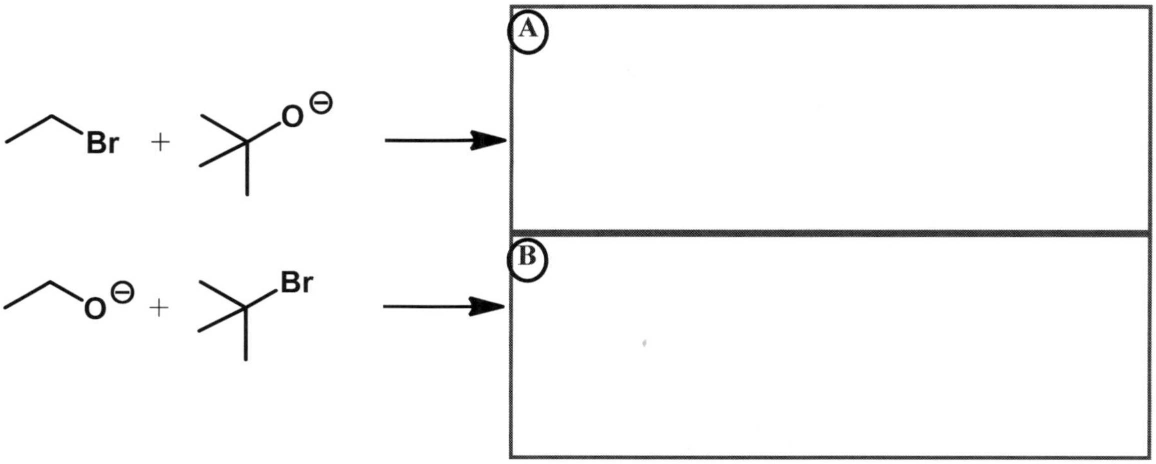


The alkyl halide should be the less-substituted group (recall: E2 favored if a tertiary alkyl halide is used, $S_N2$ favored for primary)

**Notes:**

# Nomenclature of Ethers

1.Common name:
Name each alkyl group attached to the oxygen; put the two in alphabetical order followed by "ether" (three words)

2. IUPAC name:
Pick the longest chain as parent. Name the shorter chain plus oxygen as a substituent (i.e., the list to the right)

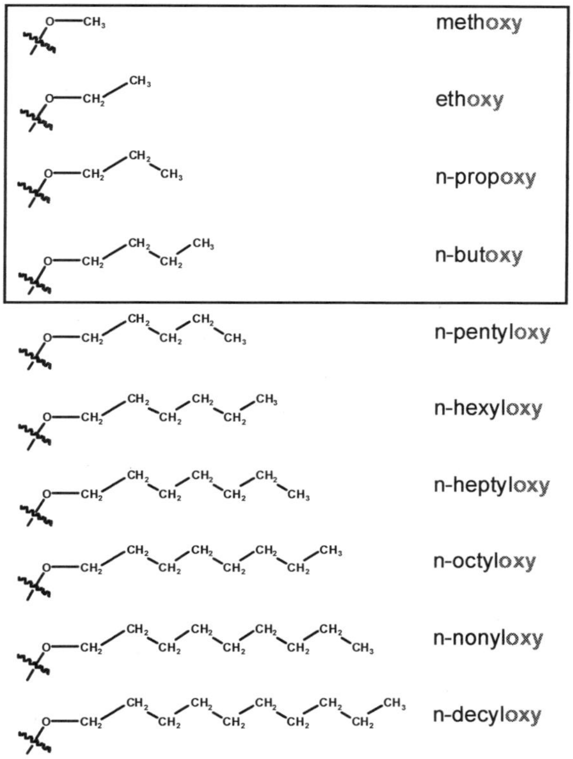


**Notes:**

# Ether Naming Examples

| | Common Name | IUPAC Name |
|---|---|---|
| 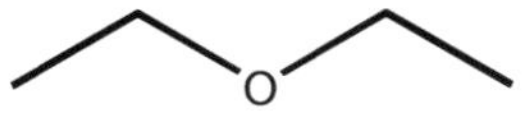 | diethyl ether | ethoxyethane |
|  | A | |
| 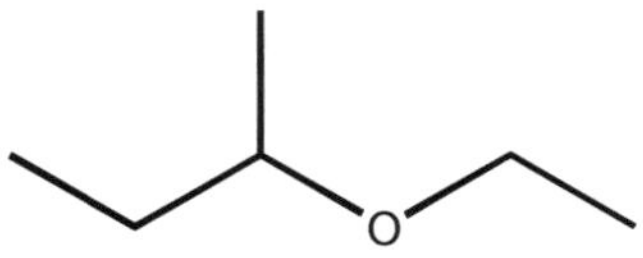 | *sec*-butyl propyl ether | 2-ethoxybutane |
| 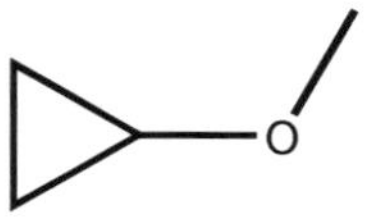 | B | |

**Notes:**

# Putting it all Together: Multistep Synthesis

## I. Retrosynthetic Analysis

**1. Identify the functional group to be made**
**2. Consider the method(s) you know to form a product containing said functional group; if more than one choice, which is better/best?**
**3. Determine what starting material you'd need to make the target utilizing the reaction you choose.**
**4. How do you make the requisite starting material? (go through cycle again until an 'allowed' starting material is unveiled.**

$\Rightarrow$ Retrosynthetic arrow; means "is synthesized from" the compound to which the arrow points.
For example:

(A)

Means:
"A is synthesized from B using reagent X."

**Notes:**

# Putting it all Together: Multistep Synthesis

Example. Indicate a sequence of reactions whereby the indicated products can be prepared from the provided starting materials in no more than three steps.

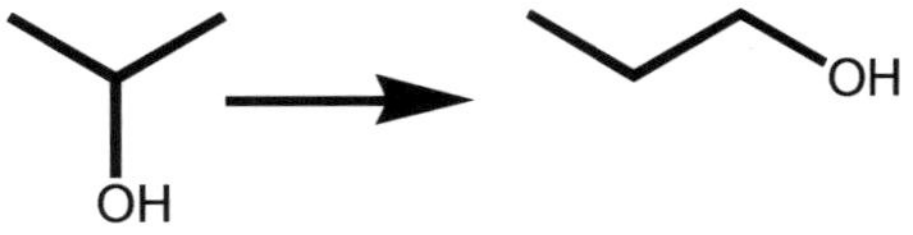

**Notes:**

# Putting it all Together: Multistep Synthesis

Example. Indicate a sequence of reactions whereby the indicated products can be prepared from the provided starting materials in no more than three steps.

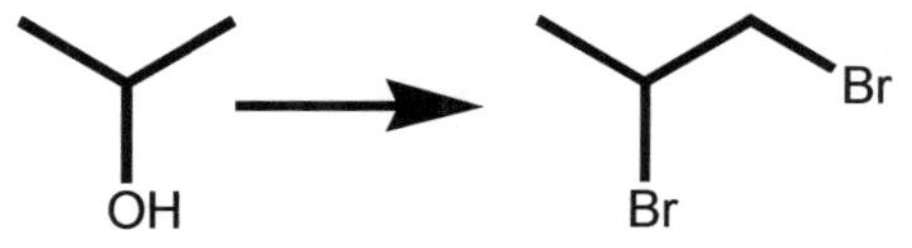


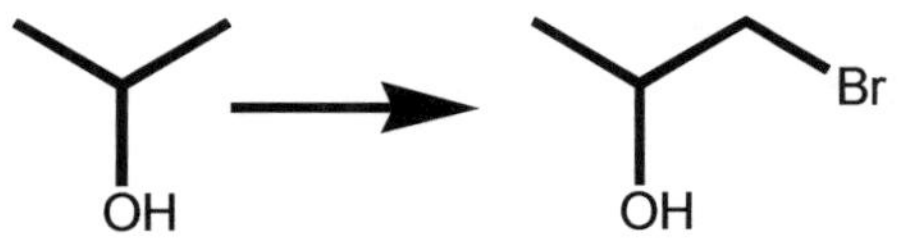


**Notes:**

# Multistep Synthesis Problem

Example. Provide a stepwise plan that could be used to prepare the target compounds. Show the product of every step. Show correct stereochemistry everywhere.

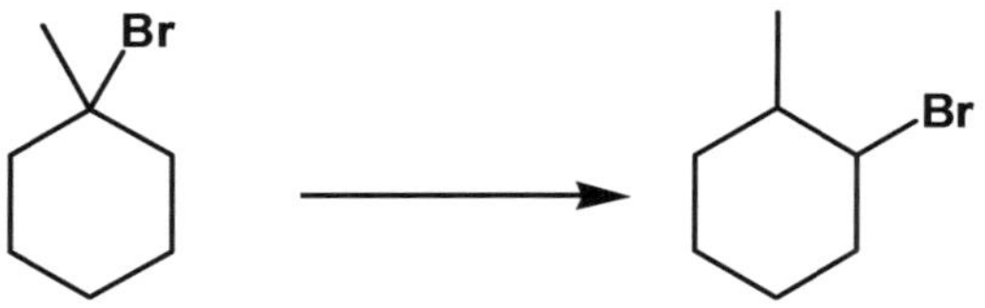


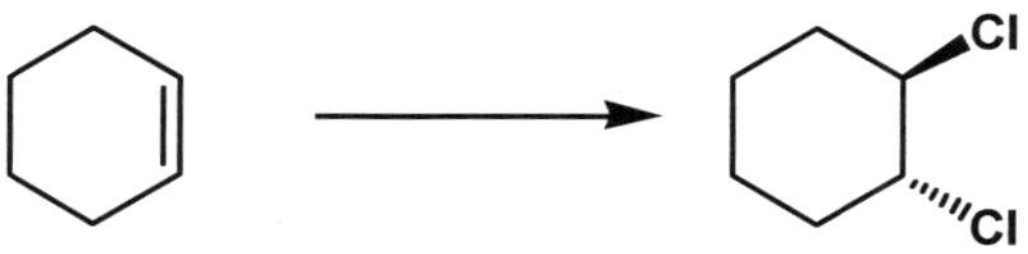


**Notes:**

# Multistep Synthesis Problem

Example. Provide a stepwise plan that could be used to prepare the target compounds. Show the product of every step. Show correct stereochemistry everywhere.

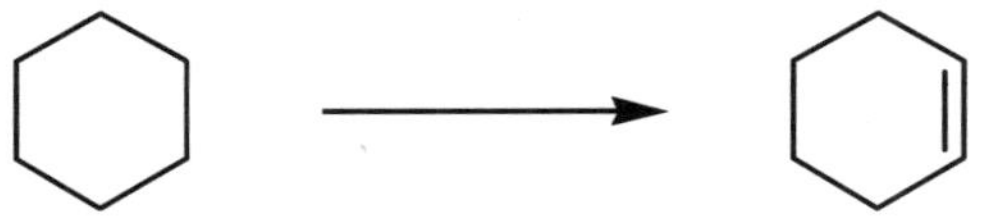

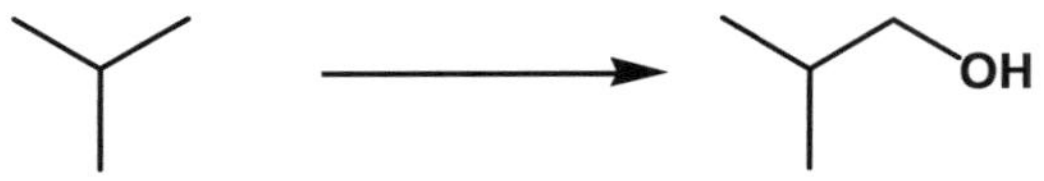


**Notes:**

# Multistep Synthesis Problem

Example: Show a reasonable route to prepare the products shown starting with acetylene or propyne and any alkyl halides. Show the product of each proposed step.

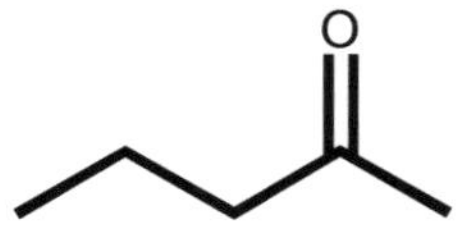

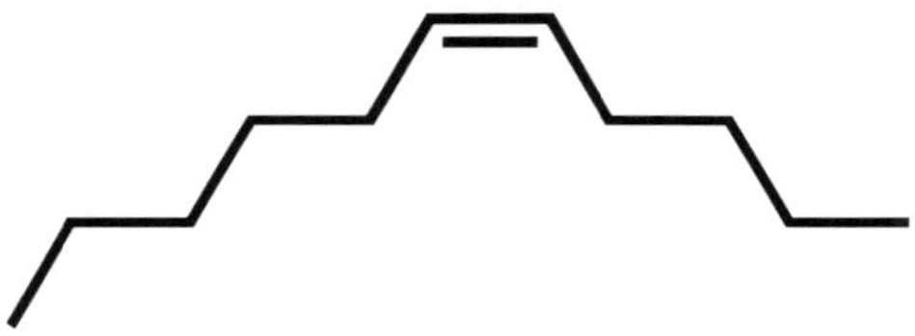

**Notes:**

# Lecture Set 10: Reactions of Alcohols and Ethers

**Suggested Reading:**

**Suggested Problems:**

# Part I: Reactions of Alcohols

What reactions will we know from Organic 1 and Organic 2 that use an alcohol as a starting material?

A. Alcohol to Alkyl Halide
   1. By reaction with HX (via $S_N1$ or $S_N2$)
   2. Using $SOCl_2$, $PX_3$ (X = Cl, Br)

B. Alcohol to Alkene (dehydration via E1)

C. Alcohol reaction with sulfonyl chlorides to make sulfonate esters

D. Alcohol reaction with acid chlorides to make esters

E. Oxidation of alcohols to make aldehydes, ketones, or carboxylic acids

**Notes:**

# Substitution Reaction: $S_N1$

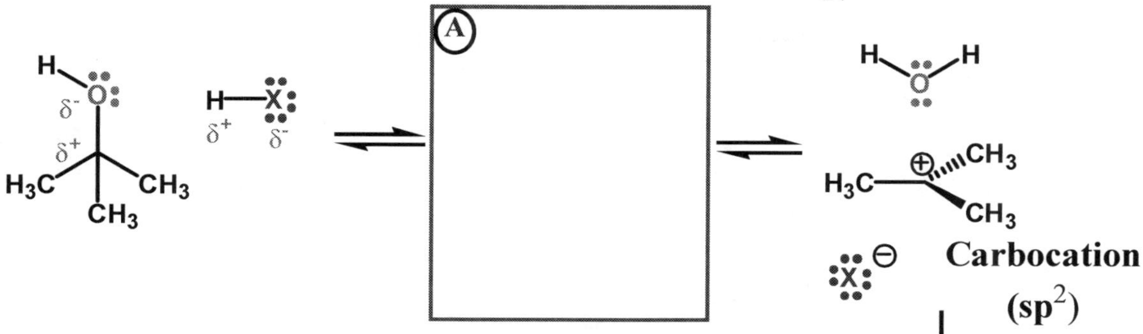


**C** **MECHANISM IS:**

**Cation can:**

1. Activate OH to make it a good leaving group
2. Group leaves
3. Anion substitutes

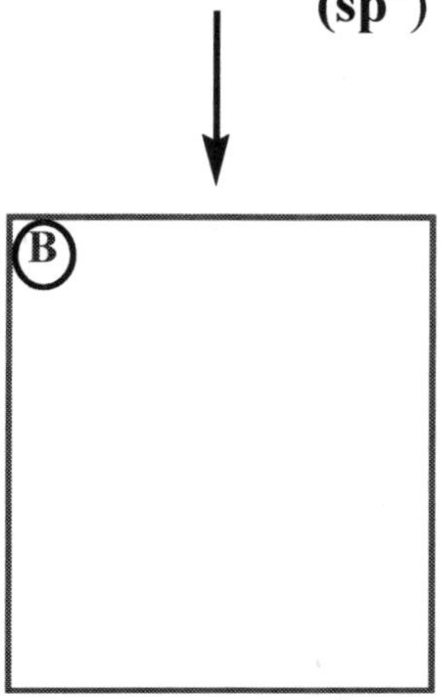


# Notes:

# The $S_N2$ Reaction Mechanism

Because (A)

reaction between *primary* alcohol and HX proceeds by a mechanism

(B)

**Primary alcohols** undergo the

(C)

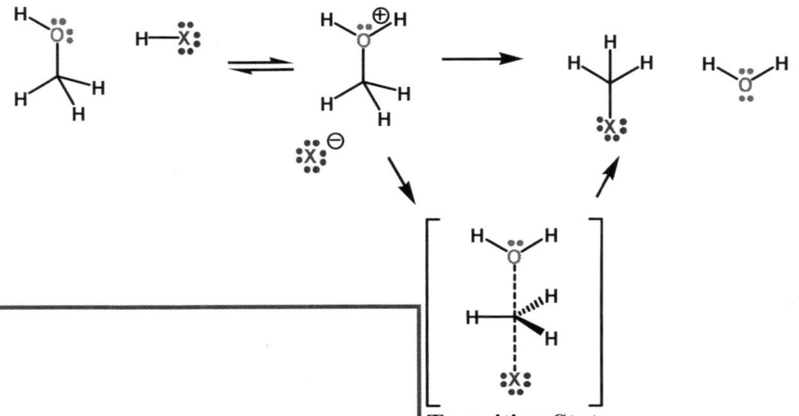


(D)

1. Activate leaving group
2. $S_N2$ reaction

(recall: dashed line is a *partial* bond, i.e., less than two electrons, so C does NOT have 10 electrons here!

**Notes:**

# Other Ways to make OH a good LG

(A)

$R\text{-}OH \xrightarrow{PX_3} R\text{-}X + X_2PO^- + H^+$

$\left( 3\ R\text{-}OH \xrightarrow{PBr_3} 3\ R\text{-}Br + PO_3^{3-} + 3H^+ \right)$

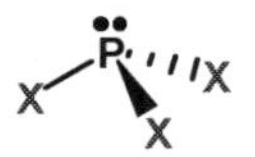


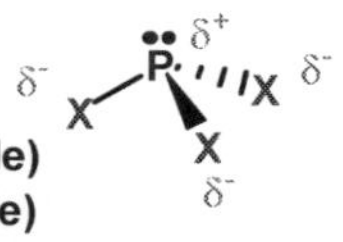


$PX_3 = PBr_3$ (phosphorus tribromide) or $PCl_3$ (phosphorus trichloride)

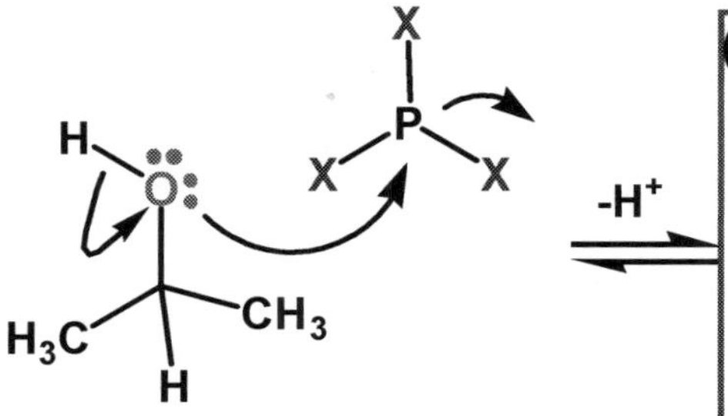


(B)

(C)

1. Activate leaving group
2. $S_N2$ reaction

Adding base drives reaction by neutralizing the acid produced.
Note: 1 mole $PX_3$ can produce 3 mol RX

**Notes:**

# Other Ways to make OH a good LG

HX is not the only reagent capable of transforming an alcohol into an alkyl halide. On the basis of the mechanism, any agent with a polar element-halogen bond that also activates OH to a good leaving group could work.

A

$$R\text{-}OH \xrightarrow{SOCl_2} R\text{-}X + SO_2 + H^+ + Cl^-$$

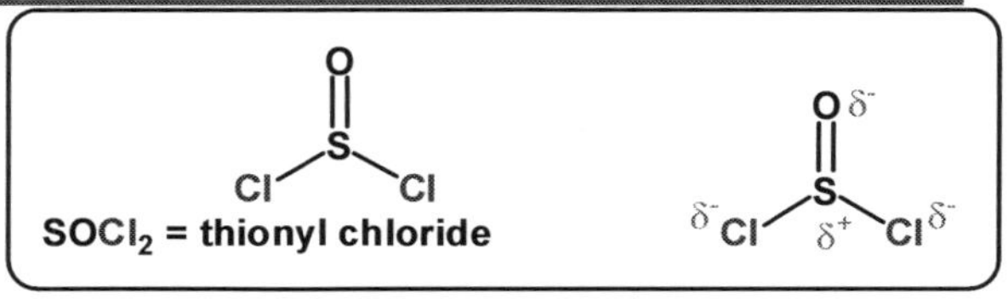


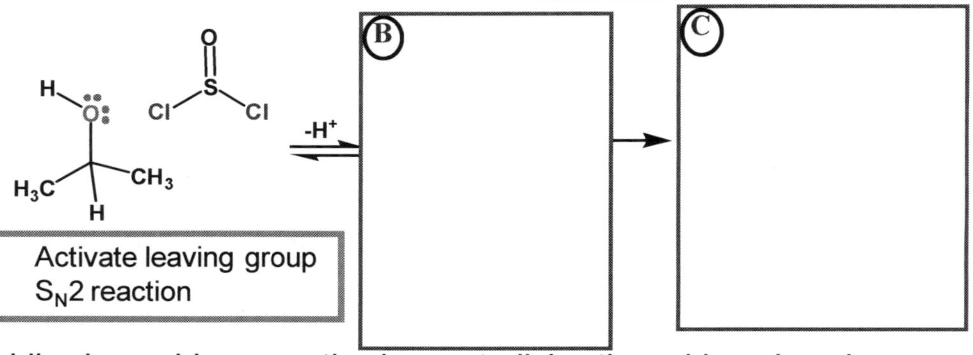


Adding base drives reaction by neutralizing the acid produced.

**Notes:**

# Other Ways to make OH a good LG

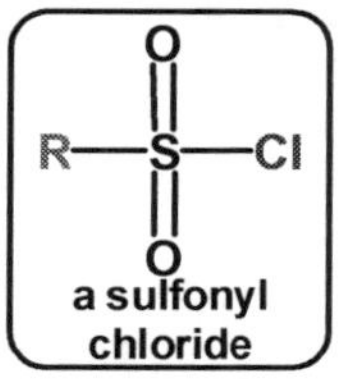


Now consider a very similar reaction where a sulfonyl chloride is reacted with an alcohol:

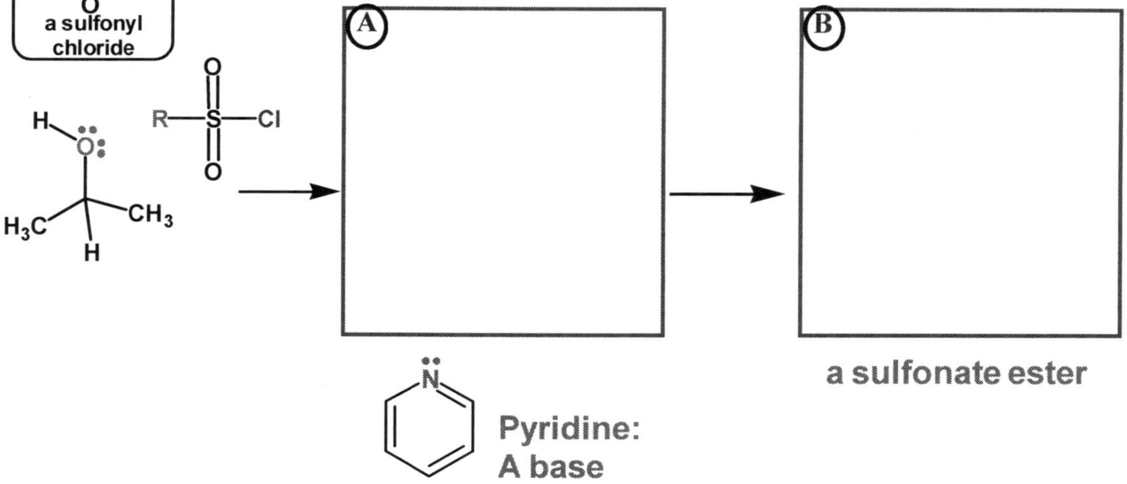


A sulfonate is an outstanding leaving group, so sulfonate esters are good starting materials for $S_N2$ reactions ...

**Notes:**

# Some Common Sulfonate Esters

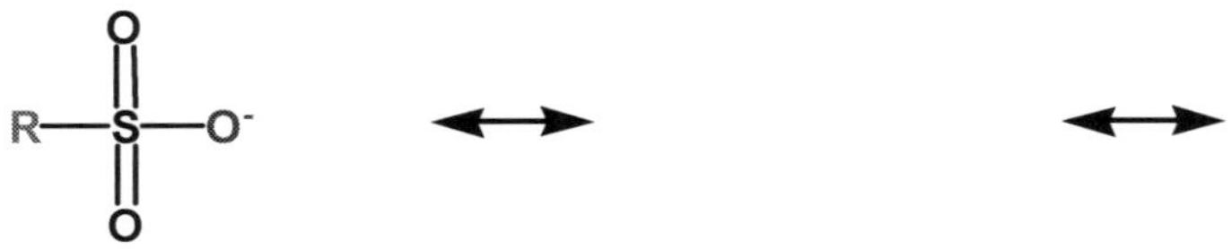


In effect, converting the hydroxide group to a sulfonate ester makes it an ideal starting material for nucleophilic substitution:

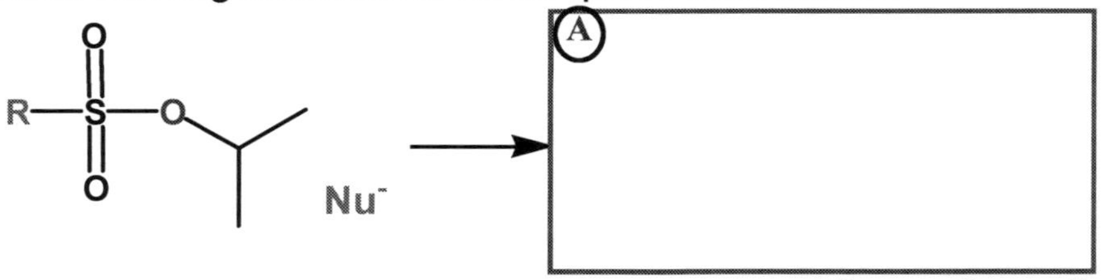


Two specific sulfonyl chlorides are often used to give particularly notable leaving groups:

*p*-toluenesulfonyl chloride

trifluoromethanesulfonyl chloride

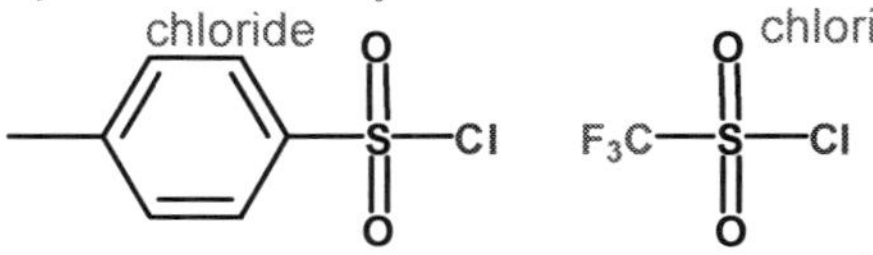


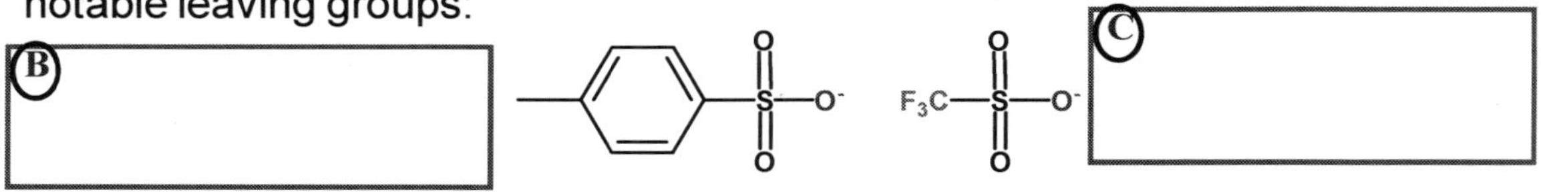


**Notes:**

# Reaction of Alcohols with Acid Chlorides

Alcohols can react with acid chlorides just as they do with thionyl chlorides!

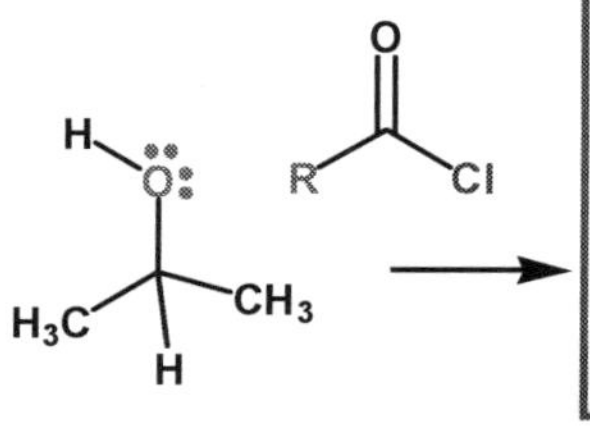


(A)

$B^-$

(base = pyridine, for example)

(B)

****BEST LG LEAVES!!!!!****

Going from **B** to **C**, note that the leaving group:

(D)

(C)

an ester

**Notes:**

# Problem: Esters and Sulfonate Esters

Example. Provide the reactant, reagent(s), or products to complete the reactions given below:

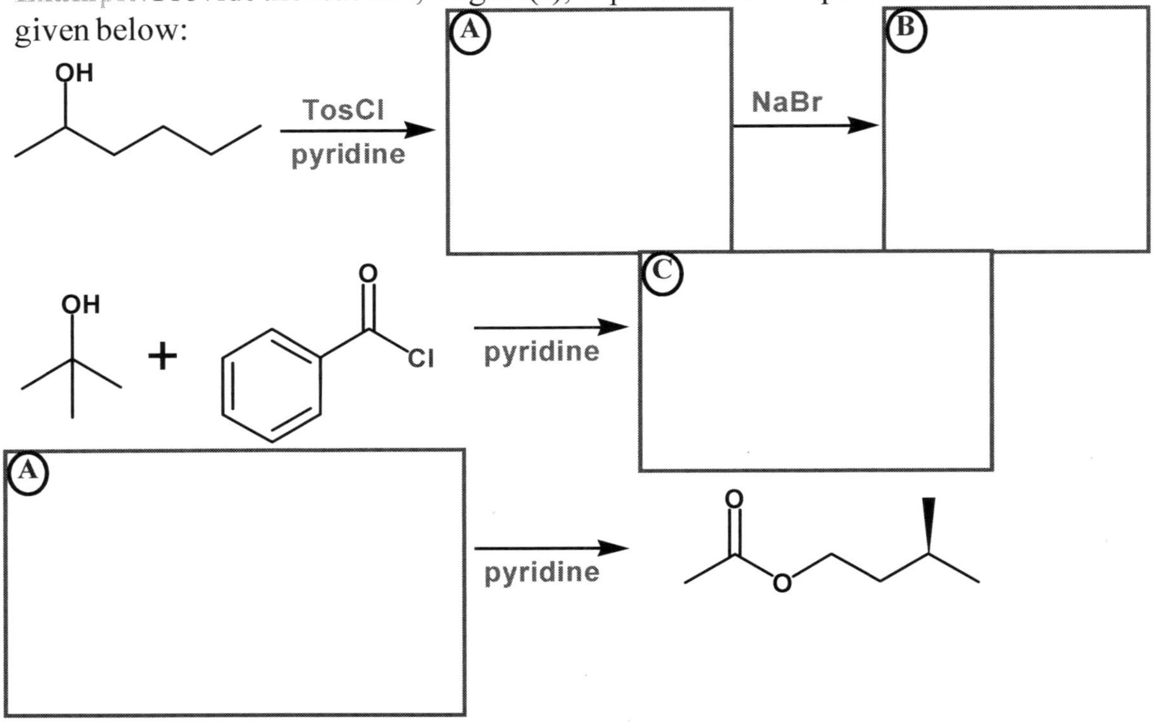


**Notes:**

# Oxidation and Reduction

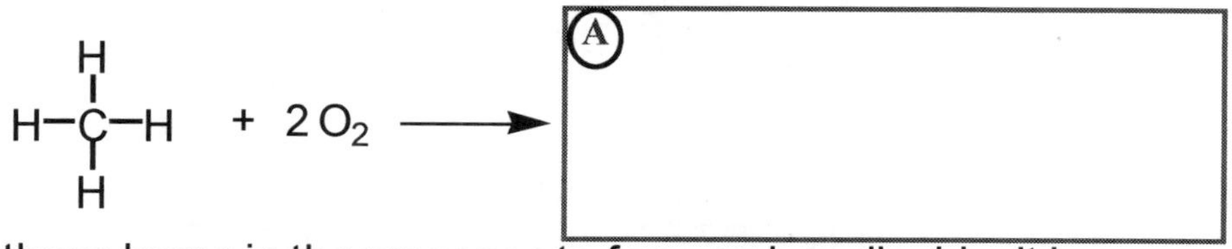


When methane burns in the presence to form carbon dioxide, it is said to undergo "oxidation". The opposite of oxidation is "reduction"

In general, the more C-H bonds that are replaced by C-O bonds in an alkane, the more oxidized the carbon atom becomes:

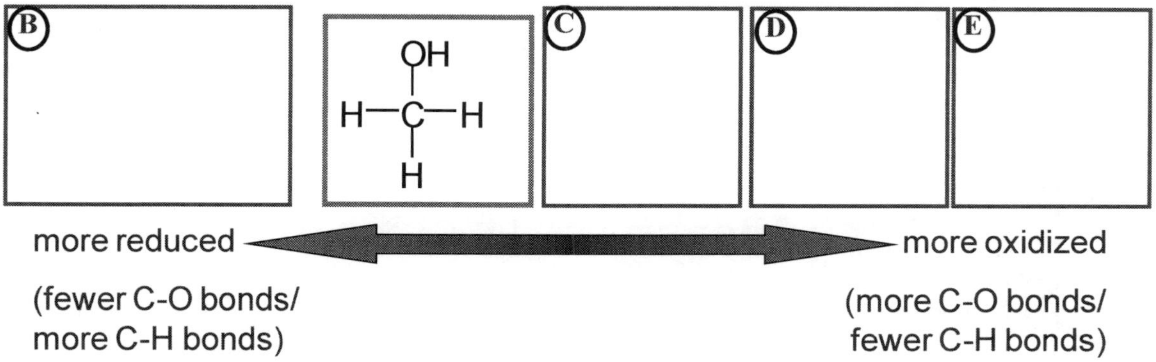


**Notes:**

# Oxidation of Alcohols

Oxidation of alcohols is a useful way to make carboxylic acids, aldehydes, and ketones. The **oxidizing agents** typically used are transition metals in which the metal is in a high oxidation state; there is thus a driving force for the metal complex to oxidize something (the metal is concomitantly reduced).

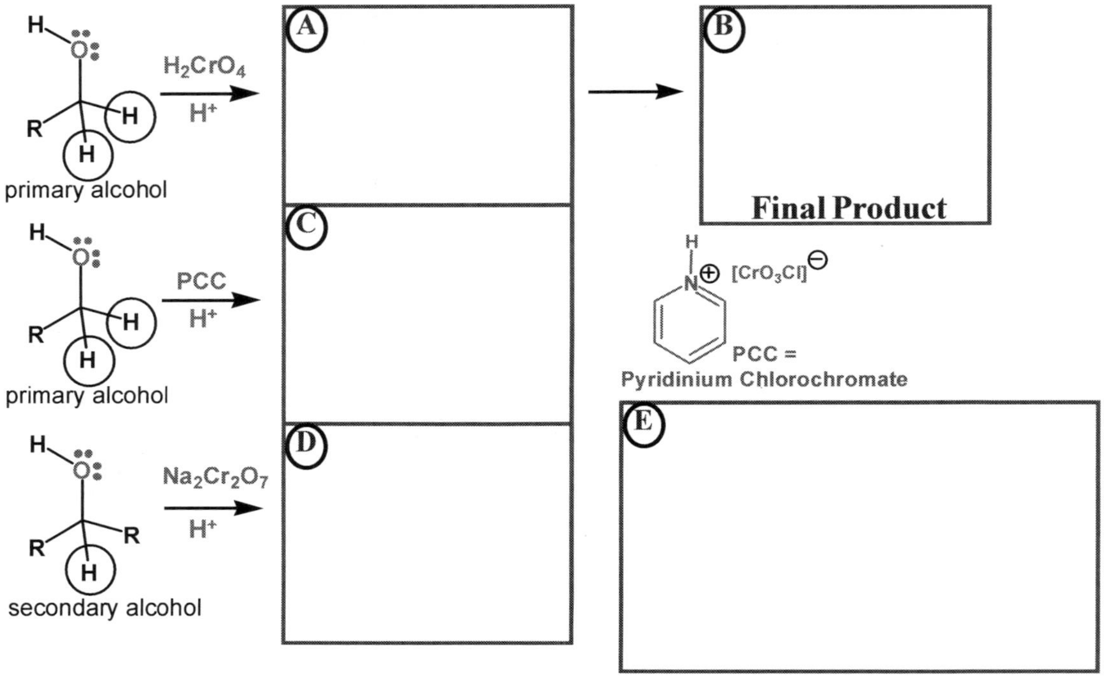


**Notes:**

# Oxidation of Alcohols

The Mechanism of alcohol oxidation is fairly simple:

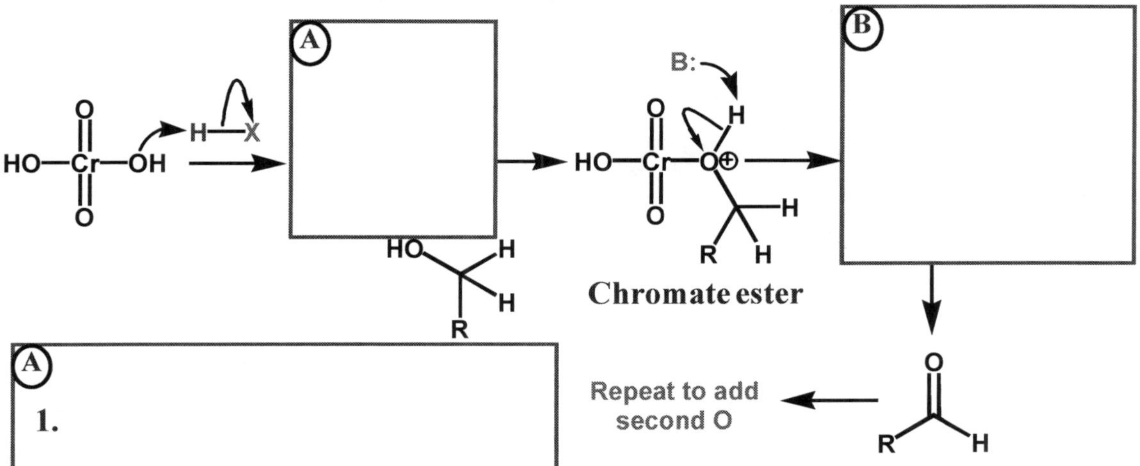


Ⓐ

1.

2.

3.

4.

**Notes:**

# Ring Opening of Epoxides

Epoxides are useful starting materials for nucleophilic substitution:

**1.**

Ⓐ

**2.**

Ⓑ

**Notes:**

# Mechanism Dictates Regioselectivity

Which carbon is attacked depends on the conditions / mechanism

(A)

(B)

**Notes:**

# Basic Conditions/Good Nucleophile

Under basic conditions in the presence of a good nucleophile, we have conditions suitable for an $S_N2$ reaction.

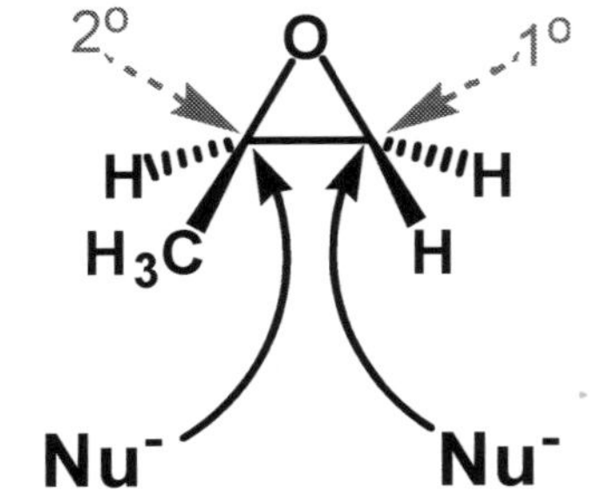


(A) The leaving group:

The nucleophile preferentially attacks:

(B)

A specific example:

**Notes:**

# Acidic Conditions

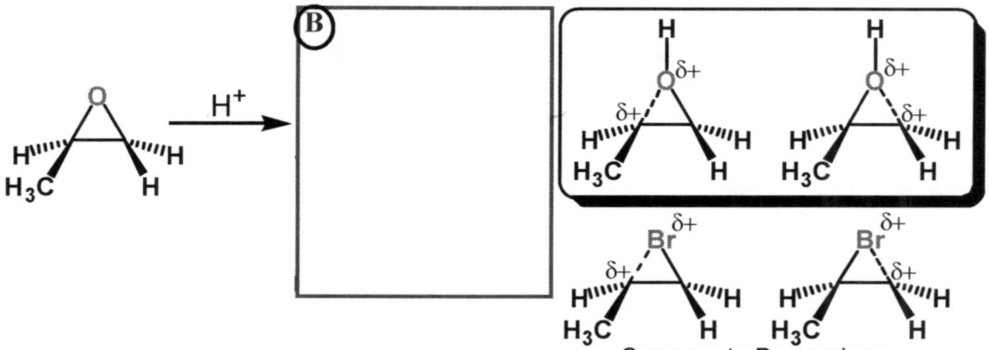


Compare to Bromonium;
we know where Nu⁻ attacks!

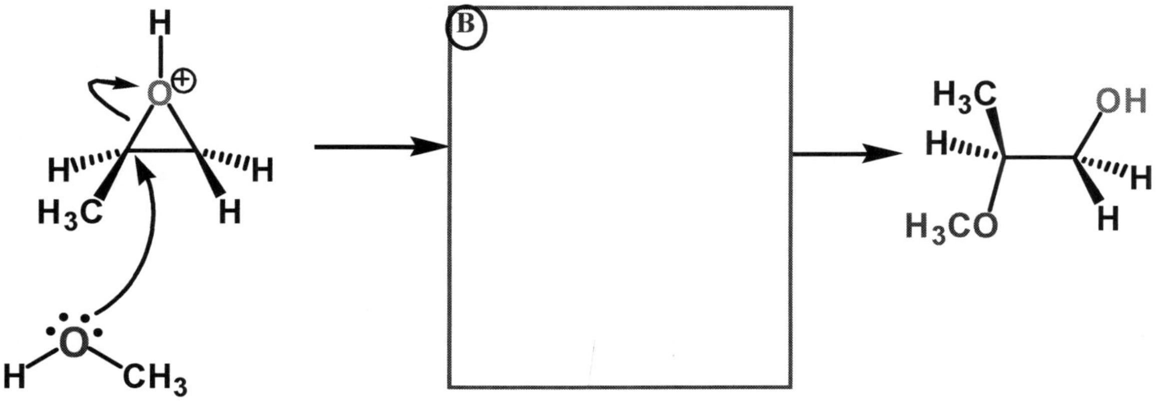


**Notes:**

# Stereochemistry

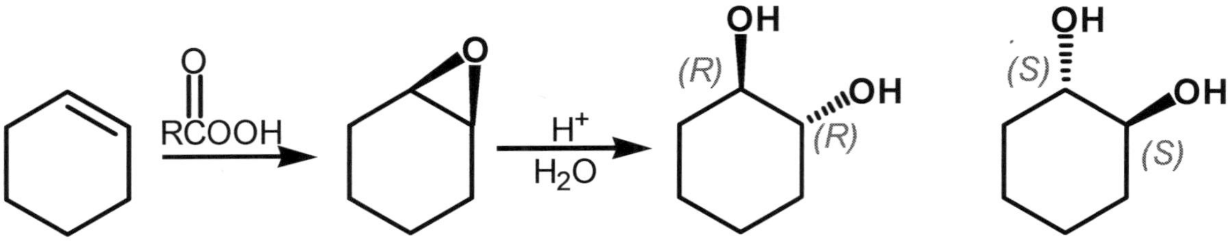


A

**Addition is:**

**Stereochemistry:**

## Notes:

# Organometallic Compounds: C–M bonds

Epoxides can be used as starting materials to add additional carbon atoms if a carbanion source is used:

A C–M (M = a metal) bond is very polar and places substantial negative charge on the carbon:

C

**Notes:**

Printed in Dunstable, United Kingdom

91342620R00201